Nary a Horse Left Standing

# Nary a Horse Left Standing

## How a Nineteenth-Century Epidemic Transformed Medicine, Animal Welfare, and the Modern City

Lee Alan Dugatkin

**The MIT Press**
Cambridge, Massachusetts | London, England

The MIT Press
Massachusetts Institute of Technology
77 Massachusetts Avenue
Cambridge, MA 02139
mitpress.mit.edu



The MIT Press would like to thank the anonymous peer reviewers who provided comments on drafts of this book. The generous work of academic experts is essential for establishing the authority and quality of our publications. We acknowledge with gratitude the contributions of these otherwise uncredited readers.

This book was set in ITC Stone Serif Std and ITC Stone Sans Std by New Best-set Typesetters Ltd. Printed and bound in the United States of America.

Library of Congress Cataloging-in-Publication Data is available.

ISBN: 978-0-262-05454-6

10 9 8 7 6 5 4 3 2 1

EU Authorised Representative: Easy Access System Europe, Mustamäe tee 50, 10621 Tallinn, Estonia | Email: gpsr.requests@easproject.com

To my dearest friend on the planet, Henry Bloom, who has been with me through thick and thin for the last half century.

# Contents

# Preface

Imagine a hypothetical strain of influenza that infects cars, vans, trucks, and trains rather than people. It's not a particularly nasty strain of mechano-flu—only a very small proportion of victims end up in the junkyard—but one that is highly transmittable. An outbreak erupts at an automobile manufacturing plant. Symptoms take a week before they're obvious, and in that time the disease spreads to every car at the plant.

Before anyone realizes what's happening, the infected cars are being shipped all over the country by trucks and trains, which themselves pick up the disease. In a matter of weeks, mechano-flu spreads everywhere. Turn the key, and all you hear is the motorized version of wheezing and coughing. Medical doctors and engineers work furiously to figure out what is happening.

During the mechano-flu epidemic people take stock of just how much of their everyday lives, as well as the lifeblood of commerce in the country, revolves around these ailing motor vehicles. No ambulances. No fire trucks. No police cars. No hearses. No garbage trucks. No cars. No buses. No trains. No Amazon deliveries—virtually no deliveries at all. No taxis, Ubers, or Lyfts. They are all convalescing. People are limited to whatever is in walking (or biking or skateboarding) distance. Factories shut down because workers can't come in, and even if they could, there is no way to get goods either into the factory or out to market. The lights on Broadway are off, as people can't make it in to the city to enjoy the theater. Movie houses and sporting areas suffer a similar fate. Price gouging kicks into high gear. Eventually, the mechano-flu runs its course, and only the occasional motor vehicle is sidelined for a few days. But in the interim, so much that we take for granted grinds to almost a complete stop.

Now, turn from the hypothetical and move the clock back 150 years. In the 1870s, horses were the lifeblood of every major city across North America. New York City, Boston, and Chicago were each powered by tens of thousands of work horses living in stables and liveries. Other cities, like Toronto, Montreal, Philadelphia, Detroit, Richmond, New Orleans, San Fransico, and so many more, had fewer horses, but the animals were no less important to every aspect of social and commercial life. Then, in the fall and winter of 1872, continuing into early 1873, an epidemic of horse flu (equine influenza) struck, so devastating that it has been dubbed the Great Epizootic—the animal equivalent of an epidemic.

# Introduction

The first cases of the Great Epizootic were recorded in the last week of September 1872 in Markham, Canada. After moving south twenty miles and grinding life to a halt in Toronto, the disease spread in all directions, including northeast to Ottawa, Montreal, Quebec City and south to Detroit and Buffalo. From Buffalo it headed east to Rochester, then Syracuse, and less than a month from when it had first been noted in Markham, it was in New York City, Boston, Chicago, and many other cities. In the newspapers, particularly early on, the outbreak of what was widely diagnosed then and almost universally accepted today as equine influenza went by many names, as people tried to sort out its cause and how it was being transmitted. Many newspapers called the disease the horse flu, others called it laryngitis, horse distemper, catarrh, catarrhal fever, the epizootic, or the epidemic. Soon epizootic made its way into the vernacular. Then the Great Epizootic.

The Great Epizootic quickly tore across Canada and the United States. Only about 2–3 percent of the animals died, but the disease spread across North America like wildfire, often following the railway routes on which horses were transported from city to city. Horses were put out of commission, wreaking havoc across cities, large and small. The epidemic seemed to come out of nowhere and struck fear into the hearts of the dwellers of America's and Canada's metropolises. One day all the horses in a city were powering urban life, and within days, hundreds, sometimes thousands, fell ill.

This was hardly the first epidemic among horses, nor was it the first to cross international borders. Historians of medicine have suggested symptoms reported more than two thousand years ago by a Greek veterinarian named Absyrtus may have been the result of equine influenza, though the extent the disease spread is not known. Hannibal's horses were attacked by some sort of flu-like disease in the marshes near Etruria in central Italy. In 791, Charlemagne's steeds suffered a similar fate. In 1299, what was an epidemic of some sort, perhaps flu, killed more than a thousand horses in Seville, Italy; in 1328, an epidemic broke out in the horses of Yemen; and then thirty years later, six thousand horses died as the result of a "plague" in England. "An epizootic [struck] among the horses of the French army, operating in Germany in 1648, which closely agrees with influenza," wrote veterinarian James Law shortly after the Great Epizootic. It "began by fever, great prostration, tears running from the eyes, and a profuse greenish mucous discharge from the nostrils. The appetite was lost and ears cold. . . . This appears to have closely followed the epidemic influenza of 1647." Law also reports epidemics, likely of horse flu, in England, France, and Germany in the mid-1830s and epidemics of both horse flu and human flu in Britain throughout the 1850s.

The North American epizootic that bore the greatest resemblance to the Great Epizootic was a devasting outbreak of glanders (also called farcy and now known to be caused by the virus *Burkholderia mallei*) among horses during the American Civil War. While there was no known cure or vaccine, military historians of the time knew of the disease, as it had plagued Constantine the Great's army, struck the horses used during the crusades, and hit the French Army during the Seven Years' War. Though it is not known for certain how or when glanders first appeared in the United States, it may have been via horses used by the British during the American Revolution.

Depending on military needs, horses were shipped in large numbers and quickly by rail often during the Civil War. Glanders has an exceptionally long incubation period of two weeks or more, and so asymptomatic but infected animals were boarded on boxcars and stockcars. When not on trains, or with a military unit, horses were often housed in giant depots, like the horse depot at Giesboro Point near Washington, DC, that could, and often did, house an astonishing thirty thousand horses. In such a depot, horses were not put in individual stalls, but rather moved about, sometimes tethered together in groups, sometimes not, in fenced off pastures and fields, one upshot of which was that the contact rates between animals was far higher than in a typical stable. Because of the high transmissibility associated with glanders, boxcars, stockcars, and depots allowed the disease to spread very quickly. Glanders wreaked wide-scale havoc in both Confederate and Federal military horses. The rail system facilitated the movement of horses infected with glanders during the Civil War, but the disease was largely a problem for military, not civilian, populations of horses (and therefore the humans who relied on them).

The Great Epizootic was different from prior epidemics, even glanders, because in the 1870s horses powered all manner of urban life and because life in large, crowded, unsanitary stables and liveries allowed horse flu to spread like a brush fire. What's more, with an incubation time ranging from twenty-four to seventy-two hours, horse flu was especially dangerous because it could, and often did, infect every horse in a stable overnight. And because some horses showed no symptoms for three days, seemingly healthy horses, who might in fact be infected, were out doing their jobs while infecting their equine brothers and sisters, or perhaps even worse, were shipped on the rails from city to city creating conditions ripe for an epidemic.

Just as the Gilded Age was commencing, and men poised to make grand fortunes were seen as captains of industry by some but robber barons by many, the Great Epizootic struck. Life in urban centers—from north, east, south, and west of North America and beyond—slowed to a crawl. Products of every sort were stranded at wharfs. The movement of food, drink, coal, and more all relied on horse power. Horses pulled streetcars, omnibuses, mail carts, fire trucks, hearses, and garbage carts. They transported people to work, to visit family and friends, to attend church services, theater performances, and more. But not during the Great Epizootic. Oxen and even humans sometimes stepped in to provide muscle power in lieu of horse power, but it was hardly enough. Monetary damage was just one cost urban centers paid. People's social lives became restricted to what was within walking distance, which meant that theaters, restaurants, baseball stadiums, operas, and circuses saw drops in attendance. Smaller cities and towns were also hit hard, though not quite as hard.[1]

But there is more to the story of the Great Epizootic than its immediate effect on life in North America's urban centers. Much more. For the Great Epizootic helped usher in three revolutions: in medicine, animal rights, and transportation. When the epidemic struck, veterinary medicine as a science barely existed in the United States, but the detailed final medical reports written on the Great Epizootic helped bolster the still-nascent germ theory of disease and dispose of a long-standing alternative theory. What's more, the epidemic provided much needed attention to early animal rights groups such as the fledgling ASPCA, which acted as guardian to sick horses forced to work even when they were victims of the epidemic: indeed, the Great Epizootic played a major role in bringing the ASPCA and the idea of animal rights to national attention. The Great Epizootic also helped usher in an era when people would rely on mechanized urban transportation rather than on living, breathing creatures to shepherd them about.

# 1
# Wheels in Our Great Social Machine

In an 1872 issue of the *Nation* sat an editorial titled "The Position of the Horse in Modern Society." "Our talk has been for so many years of the railroad and steamboat and telegraph, as the great 'agents of progress,'" the *Nation*'s editors told a society swept up in the technical revolutions of the age, "that we have come almost totally to overlook the fact that our dependence on the horse has grown almost *pari passu* [in equal step] with our dependence on steam."

The movement of people along the rails and rivers, as well as information over the telegraph lines, the *Nation* proclaimed, had strengthened, not weakened, our reliance on horses. True, the editors noted, "We have covered the ocean with great steamers," but "they can neither load nor discharge their cargoes without horses." Yes, "We have collected at the mouths of our great rivers and at the intersections of our railroads vast bodies of people," but "we have left them wholly dependent for their intramural travel and for their popular supplies of food and clothing on horses." What's more, lest readers forget, "We have within the last few years made horse labor an almost essential condition of the protection of our great cities from fire."

At the cusp of the Great Epizootic, large cities in the United States and Canada were powered by horses. To begin to appreciate the chaos caused by the Great Epizootic requires understanding how horses came to be the lifeblood of urban transportation in North America by 1872. The term "horsepower" was coined by James Watt, of steam engine fame, who after doing some experiments with dray horses, defined it as the amount of power needed to move thirty-three thousand pounds a distance of one foot in one minute. To produce horsepower, horses themselves needed to be powered and there was no lack of effort put in by many to calculate the optimal feeding and working schedule for horses, as well as mules, donkeys, oxen, and other beasts of burden. In his book *The Animal as a Machine and a Prime Motor, and the Laws of Energetics*, mechanical engineer Robert Thurston, who at various times had served as assistant professor of natural and experimental philosophy at the United States Naval Academy at Annapolis, professor of mechanical engineering at the Stevens Institute of Technology, and president of the American Society of Mechanical Engineers, wrote of "prime motors" as "machines by means of which man diverts energy from natural channels and compels it to do his own work." As far as animals went, Thurston argued, horses were prime motors extraordinaire.

Despite our reliance on these living, breathing, prime motors, "most of us," the editors of the *Nation* proclaimed, "have well-nigh forgotten that the horse was an animal like ourselves liable to pains and aches and death. . . . We have come to think of him as a machine, on whose endurance we could calculate as on that of an engine, and for whose mortality we could make ample allowance in our business under the head of 'wear and tear.'"

The idea that horses, indeed all animals, were machines is most famously associated with René Descartes's mechanistic physiology in which beasts—nonhuman animals—lack sentience and self-awareness and so are a kind of living automata. The *Nation* thought otherwise. There was indeed a horse/machine analogy to be made, the journal's editor's suggested, just not Descartes's analogy, but rather that "[horses] are wheels in our great social machine."

That great social machine was operating at full speed in North America's urban centers. There, where the *Nation* decried that "the working horse is treated worse than a steam-engine or sewing-machine," one of the primary occupations of the horse was not just to move people—a job it had done admirably for thousands of years—but also to move people *en masse*. Transportation en masse was critical in the urban centers across North America, where population growth had exploded over the decades leading up to the 1870s. In the early 1830s, Toronto's population hovered around ten thousand; shortly before the Great Epizootic that number was nearly six times higher. In the 1830s, New York City was home to 202,589 people. Over the next decade its population grew by 68 percent, then 64.9 percent between 1840 and 1850, 57.8 percent between 1850 and 1860, and 16 percent between 1860 and 1870, when New York City's population sat at 942,292. During the same period, Boston's population grew from 61,392 to 250,526 and from 1840 to 1870 the population of Chicago—which was not even incorporated as a city until 1837—skyrocketed an astonishing sixty-five-fold, from 4,470 to 298,977. All those people needed to be moved from place to place within cities, as well as between those cities and the neighborhoods that surrounded them: neighborhoods that were growing at an even faster rate than the cities themselves.[1]

The omnibus, the first major breakthrough in horse-powered urban mass transportation, modified the body plan of the stagecoach, which largely moved people *between* cities, for use in intracity transportation. Parisians were the first to ride on omnibuses—"buses for all"—thanks to philosopher Blaise Pascal, who in 1662 convinced Louis XIV to fund the building of seven such horse-drawn vehicles.

Pascal's omnibus held more people than carriages, but only slightly more. What made it different was that it ran on regular routes, on a fixed schedule, and did so for a relatively nominal fee. But within a decade Pascal's enterprise dissolved, largely because wealthy noblemen and businessmen were uncomfortable with the notion of mixing with laborers, especially in close quarters, and soon saw to it they alone would be permitted to ride.

The omnibus rose from the ashes in France in 1828. Like its predecessor, it ran on fixed schedules and charged reasonable prices, but now the buses were designed to seat up to two dozen people. Within a year there were ten companies running hundreds of horse-drawn omnibuses in Paris alone. London followed suit in 1832 when Parliament chartered omnibuses and within a few years, more than six hundred seating up to twenty people each were navigating the crowded streets of London.

Before Abram Bower and John Stephenson introduced the first true American omnibus to New York City in the spring of 1831, the only pay-for-hire alternative to walking the streets of that city (or any large city in the United States) was the hackney coach, or hack, an early 1800s version of a taxicab that used a horse, or a pair of horses, to power a covered coach that held up to six people. And for most New Yorkers, who might make a dollar or so a day for their labor, a hack wasn't really an option at all because the fare, which ran from twenty-five cents (two bits) to thirty-seven and a half cents (three bits) for any distance up to a mile and went up from there, priced out all but the wealthiest riders (figure 1.1).

After experimenting with omnibus models they named the *Accommodation* and the *Sociable*, both of which still had a markedly stagecoach look and feel to them, Bower and Stephenson launched a new model. Lined with windows along their sides, these vehicles, whose routes covered portions of Lower Manhattan, could seat between twelve and twenty-eight passengers. Because omnibuses weighed over a ton and carried another ton or two in passenger

**Figure 1.1**
One of Bower's early omnibuses. From E. Stratton, *The World on Wheels* (New York: Published by the author, 1878), 438.

weight, they were typically powered by three or four horses, each of which tipped the scales at about a thousand pounds. To address this, Stephenson added new elliptical springs to cushion the blow of this three- to four-ton behemoth traversing bumpy cobblestone.

As on the *Sociable*, in the newer model a driver sat on a platform at the front and passengers boarded and departed from the rear, rather than through side doors, which would have had to have been opened and closed at every stop. Such time-minimizing innovations were needed to speed up the process, as early on omnibuses tended to move at a rather lumbering two or three miles an hour. As in Paris, they traversed fixed routes, and followed a set schedule to the extent possible in an already bustling New York City. In winter, if the snow was sufficient, drivers would swap the spoked wheels for runners called bobs, turning their vehicle into a sleigh.

Omnibuses were often decorated with flags and placards, and given monikers such as the *Knickerbocker*, the *Empire*, the *Waverley*, or for the companies with a more patriotic bend, the *George Washington*, the *Lady Washington*, the *Benjamin Franklin*, and the *Dewitt Clinton* (who had served as the mayor of New York City thrice, and as governor of the state). The fare for a single ride was usually twelve and a half cents (one bit), significantly less than what riding a hack cost. With the omnibus driver on a platform at the front, facing the road, the fare was sometimes collected by a young boy at the rear entrance, but more often placed in a locked box up near the driver, who would provide change when need be. Once on board, a passenger who wished to leave a Bower omnibus would pull on a leather strap that ran from the back of the bus near the door up to the front, where it was attached to the driver's ankle. Soon this was replaced by an ornate bell in the center of the coach, which passengers could ring by tugging on brass pulleys when they wished to exit.

Bower's omnibus was novel enough that it was hard to miss. The year of its maiden voyage, it appeared in Hugh Reinagle's painting *View of St. Paul's Church and the Broadway Stages, N.Y.*, which was being raffled at the Bank Coffee House on the corner of William and Pine Streets, at the rear of the Bank of New York. Later that same year, competing omnibus lines—such as The Kip and Brown Accommodation Line, whose *White Lady* model, named after a character in a Walter Scott novel, had a "handsome lady painted on the center"—began operation. The omnibuses of the Greenwich, Drydock, Bowery, and Broadway lines were "whirled through the streets by handsome and spirited horses." Access to omnibus routes soon became a selling point used by real estate brokers: "[This] property is calculated to become of great value," read an advertisement in the *New York Evening Post*, not only because it "was adjacent to the dry dock," but also because the roads were "regulated and paved . . . and about 20 omnibuses are now running from this point to Wall Street."

Newspapers of the day were, understandably, fascinated with this new mode of mass transit. Whenever Bower introduced a new omnibus model, reporters were there to announce it: "The size of this [new] spacious vehicle is so great, that at least twenty full grown persons may very conveniently occupy it at once," wrote the *New York Evening Post* on August 11, 1831. As to the two rows of seats, which ran along the side of the omnibus, rather than across its breadth as in stagecoaches and hacks, "the space between the seats is more than a yard and from one side of the carriage to the other nearly six feet." In a bit of free advertising for Bower, the paper also informed readers that "the floor is level and covered with handsome oil cloth; the seats are neatly cushioned and the trimming of the whole interior of good quality and neatly put up. . . . The exterior . . . has been decorated with much pains and skill, so that it is in this respect, as well as in its interior accommodations and finish, very superior to the ordinary Broadway stages."

Children too were fascinated by these strange new buses, but more often than not, lacking the money for a ticket, they simply jumped aboard. "Very scarcely an omnibus passes through Broadway the steps of which are not at all times loaded with two or three little urchins," one newspaper wrote a cold winter day in 1831, who are "jumping off, or being thrown off whenever the vehicle stops."

A gentlemen or lady strolling the streets of New York City in the 1830s became accustomed to seeing would-be passengers wave down a horse-powered omnibus and board. They'd also witness intense competition to land the fare of a hailing passenger. The *New York Journal of Commerce* describes how, as a result of the race to get passengers to board *their* vehicle, "the character of the . . . drivers has become brutal and dangerous in the highest degree." Drivers threw caution to the wind as "they race[ed] up and down Broadway . . . with the utmost fury . . . a hackney-coach (!) was crushed between two of them the day before," the paper decried, "[and] it is but a few days since we published the account of a physician being run down near the same spot, his gig ruined, and his horse nearly so, and his own life placed in the most imminent hazard." The *New York Journal of Commerce* demanded that action be taken to curb "a ferocious spirit [that] appears to have taken possession of the drivers, which defies law and delights in destruction."[2]

Other American cities were watching the growth of the horse-powered omnibus culture in New York City. Not long after Bower's models were coursing through Manhattan, James Boxall started Philadelphia's first omnibus service, which had an hourly circuit along Chestnut Street to the Merchants' Coffee House and Palace of Exchange. Edward Deschamps soon followed suit with a competing line. Boston's first omnibus line opened in 1834, quickly replacing its hourly stagecoach service that had moved people from downtown to Roxbury, Charlestown, and South Boston. Chicago, which was still in its early growth stage in the late 1830s, didn't see omnibus service until the early 1850s, when Franklin Parmelee opened a line that for a fare of five cents transported people between railroad stations on the outskirts of the city to hotels downtown. John Frink soon had a competing line in place.

Once city dwellers got a taste for omnibuses, they wanted more, and there were plenty of companies that were happy to oblige. In New York City, 245 omnibus licenses were granted in 1846, in 1850 that number jumped to 425, and in 1860, 536 omnibus licenses were issued. Many of those vehicles were servicing the area around City Hall, but as businesses and residences began to move north, toward what is today midtown Manhattan, omnibus routes began to spring up there as well. The 1853 edition of *The Strangers Handbook for the City of New York* informed both New Yorkers and visitors that there was an average of 13,420 omnibus trips a day, with an estimated 120,000 tickets sold per day. Assuming that most people traveled round trip, and so purchased two tickets, this translates into horses and drivers shepherding about sixty thousand passengers a day to their ultimate destinations.

The growth of the omnibus culture was not restricted to New York City. By 1848, Philadelphia had 130 omnibuses running eighteen different routes and by 1854, those numbers had soared to 320 vehicles covering thirty routes. In Chicago, the growth of omnibus lines was largely tied to intra-city commutes between railroad stations: within a few years after Parmelee's first line, Chicago was home to eighteen omnibus companies that ran more than four hundred trips a day. In time, Parmelee bought out his competition, as well as several hotels along their routes, creating an early transportation monopoly. In Boston, the growth of omnibus lines was integrally connected with the growth of Roxbury, West Roxbury, and Dorchester, all south of Boston proper, and all eventually annexed by the city. Omnibuses allowed businessmen and their families to work in Boston proper but live in what today we call the suburbs. Though not officially annexed to Boston until 1870, property owners in Roxbury were using omnibus access as a selling point much earlier: "a suite of rooms, eligibly and very pleasantly situated in Roxbury," read an 1849 advertisement printed in the *Boston Evening Transcript*, "their proximity to Boston being on the omnibus route renders them very convenient and agreeable."

A combination of improvements in urban planning, design, mass production, and horse breeding helped foster explosive growth in the number of omnibuses in large cities. In the late 1850s, Henry Burden, president of the Troy Iron and Nail Works, devised a means to mass- produce horseshoes, so that horses pulling omnibuses (and other vehicles) could be quickly and properly shod in new shoes as the old ones wore away. Cities paved streets with stone block pavers that allowed calks—cleat-like bits on the bottom of horseshoes—to get better traction. Sidewalks, a relatively rare sight before the 1850s, became commonplace, separating pedestrians from vehicles, including omnibuses. Those omnibuses were now being designed in more efficient ways, reducing their weight and so increasing the speed at which horses could pull them along. And the horses themselves were being biologically engineered—crossbred with larger, stronger, European draft horses—to pull those lighter wagons even faster.[3]

Omnibuses were more than just a means of transport. They were a microcosm of a city, where men and women of many stripes mixed in just the manner that wealthy noblemen and businessmen in

Blaise Pascal's time worried they might. And everyone—from denizens of a city to visitors from abroad—wrote of their experiences and adventures aboard them.

In his colorful 1849 book *New York in Slices/By an Experienced Carver*, reporter and litterateur George Foster tells his readers: "There is plenty of philosophy in omnibus-riding, if you have but the wit to find it out. The omnibus is an excellent school for studying human nature, besides being . . . an almost sovereign remedy for the dyspepsia." Which philosophers were teaching depended on the time of day. Writes Foster, "While the fresh sunlight streams in rosy coolness, . . . the omnibus is chiefly occupied by junior clerks with big iron keys in their hands, or laborers with tin kettles between their feet, on their way to their downtown avocations. All is perfect equality in the early omnibus." As the morning unfolds, Foster tells his readers, "a new school of philosophers take the stage," when "the sleek and rotund burghers . . . stand imposingly halfway down their palace-steps, and wave their hand with ludicrously majestic motion as the inevitable vehicle heaves in sight!" Diarist George Templeton Strong made note that on mornings when political conventions were in town, he witnessed omnibuses full of "hirsute, porky creatures, in linen 'dusters,' with badges in their buttonholes," advising one another that to prepare for the endless debate and bickering they were in store for, they might "stop at the New York Hotel and licker up."

While the morning saw largely males on board, as the afternoon approached, "the fluttering morning robes and thick green veils of Gotham's fair wives and daughters, on their various errands," boarded. "The strap is delicately pressed," Foster writes, "by the smallest hand in the world, in a ravishing brown thread glove, that clings in pure love to the exquisite proportions it does not conceal." Women's time on the omnibus was short, as they soon vanished, as Foster recounts it, "upon their various errands—one to visit her milliner or dressmaker . . . another to the dentist, to work concealment of the ravages of time and hard tooth."

As the afternoon progressed, the countercurrent rolled in and "soon the brokers and merchants . . . begin their homeward journey . . . [and] are followed by a gradually increasing crowd of all occupations, until at sundown it becomes a throng, and you cannot procure a seat—not even a standing-place on the steps . . . for love nor money." But the day was not over for the omnibuses, their drivers, and the horses that powered them, for now they become a vehicle of leisure, "drawn up in extempore barricades across the street at the doors of the theatres and concerts, patiently waiting," where "they linger far into the night, and at length straggle along up Broadway, one by one, keeping a keen look-out for stragglers on foot . . . and making the midnight ring at every corner with their 'Broadway—right up!' and the sharp crackle of the whip" (figure 1.2).

Some riders made it known that a trip on an omnibus could, depending on circumstances, be anything but pleasant, that this moving microcosm degenerates into "perfect bedlam . . . the discomforts, inconveniences and annoyances of a trip in one of these vehicles . . . [is] almost intolerable." The solution, one editorial in the *New York Times* noted, was straightforward: "Reason and right . . . appear to point out this rule: that he from his condition or his behavior who is a serious annoyance to others is to be refused the opportunity to annoy them. A person who by drunkenness, filth, or disease, or by an offensive behavior, such as smoking, or violence . . . offensive speech, such as ribaldry, profanity, obscenity . . . makes his fellow passengers uncomfortable, should be removed from the public conveyance."

All in all, riders generally took the bad with the good. "No regular omnibus-rider is without his established list of complaints against whatever line he happens to be patronising," Foster writes in *New York in Slices*. "The *Knickerbocker* creeps, the *Empire* races, the *Waverley* always takes in a regiment of children . . . [but] The Grumbler . . . will always grumble." Grumbling and complaining is part of life in New York City, Foster continues, "but how few of us, who save

**Figure 1.2**
Riders on a New York City omnibus. From G. Foster, *New York in Slices/By an Experienced Carver* (New York: W. S. Burgess, 1849), 64.

annually so many hundred furlongs of weary walking, and so many pounds of costly sole-leather by the omnibus, ever pause to think of what a blessing we are enjoying, or to ask ourselves the unanswerable question—how should we get along without them?"

Riding an omnibus in New York City was a memorable event even for people who had the opportunity to ride on one in London. When British scientist Sir Charles Lyell visited New York in 1846, omnibuses were a symbol of the wild, explosive growth taking place in the city. When Lyell was in New York City in May 1846, he looked about in wonder: "Five months since we were last in this metropolis, whole streets had been built, and several squares finished in the northern or fashionable end of the town," he wrote in *A Second Visit to the United States*. "Hence there is a constant increase of omnibuses passing through Broadway and other streets running north and south. Groups of twelve of these vehicles may be seen at one time." A few years earlier, when Charles Dickens visited Manhattan, he felt awash in such vehicles. "No stint of omnibuses here!," he wrote in

*American Notes for General Circulation*. "Great wooden ark[s]," transporting "a score or two of people" were everywhere.[4]

In the 1850s, these great wooden arks that dominated the streets of New York City, Boston, Chicago, and more, faced their first serious competition for passengers in the form of streetcars, sometimes called horsecars or horse railroads. The first streetcar line in New York City was built decades earlier, in 1832, with cars designed by John Stephenson who had worked with Bower on the early omnibuses. The birth of this early streetcar line was tied to the New York and Harlem Railroad, which had been chartered by the state legislature on April 25, 1831, with the ultimate goal of providing service along the Hudson River from the state capital in Albany, through Harlem, down to the lower tip of Manhattan. The railway charter allowed for the "transportation of persons and property . . . by the power and force of steam, of animals, or of any mechanical or other power, or of any combination of them which the said company may choose to employ." The earliest tracks laid were on Fourteenth Street, parallel to the Bowery, down to Price Street, and though the plan was for steam-powered railways to run on the tracks soon to be built up to Harlem and Albany, horses regularly pulled the cars on the original tracks within New York City: with that the urban streetcar was born.

The driving force that put the New York and Harlem Railroad into place was John Mason, the second president of Chemical Bank, and Stephenson paid him due homage, naming the first streetcar the *John Mason*. With its doors on the sides, the *John Mason* looked more like a stagecoach—or more precisely a string of stagecoaches connected to one another—than an omnibus. But its metal wheels set it apart, riding along track that was attached to street blocks, so that the rails lay above the street itself. And riding along tracks, rather than on bumpy paving stones, meant the pulling efficiency of horse-powered streetcars was up to four times greater than that

**Figure 1.3**
One of John Mason's early streetcars. From E. Stratton, *The World on Wheels* (New York: Published by the author, 1878).

of omnibuses, making the streetcar a much more efficient vehicle that could charge lower fares: often tickets cost only half a bit (6¼ cents), far cheaper than omnibuses fares (figure 1.3).

On November 14, 1832, powered by four horses and with quilted seats that could hold up to ten passengers in each of three compartments, the *John Mason* and a second streetcar made their initial runs along the Bowery. Gentlemen and ladies in their finest clothes lined the streets as the *John Mason*, driven by one "Lank" Odell, and seating Mayor Walter Bowne and the members of the city's Common Council, rode by in the first car and officials of the New York and Harlem Railroad, captained by a local hack driver, his name lost to history, were transported in the second. Though there was a bit of a snafu along the way—the driver of the second car pulled on the reins rather than applying brakes by pulling a nearby goose-necked handle on a vertical staff—the launch was celebrated in great fashion at a dinner celebration that evening at City Hall, with champagne flowing freely. Soon New York and Harlem Railroad streetcars were running Monday to Saturday, from 9:00 a.m. to 5:00 p.m., the owners adding more track, more cars, and horses to power them as routes expanded. The New York and Harlem Railroad purchased six lots at the corner of Twenty-Fifth Street and Fourth Avenue for stables and a depot for horsecars and steam railroad cars, and built a luxury hotel—Prospect Hall, under the directorship of George Nowlan—near that depot.

Though the New York and Harlem Railroad was off to a flying start, and indeed the steam rails that were soon traveling between Albany and New York City thrived over the following decades, the growth of the horsecar industry lagged behind, and did not come into full bloom until the early 1850s. The delay was, in part, caused by the way street railway tracks were laid. Because they were set above rather than flush with streets, they made crossing a bustling Manhattan street all the more difficult for pedestrians. Raised rails also caused havoc for omnibus and hack drivers as they attempted to navigate the streets. Those drivers, though, had their means to retaliate: "The daily practice of carriages and private teams to take possession of the track," the *Brooklyn Daily Eagle* wrote, "and purposely keep in front at such a snail-like pace as to seriously annoy the drivers and incommode the passengers in the [street]cars" never ceased to make the point.

Alphonse Loubat, a French inventor and engineer (and oenophile) who moved to Brooklyn, New York, came up with a solution to the raised rail problem on the streets of New York City. His girder rails were flat to the street surface and laid in when new street paving was poured. Just like that, there were no more horses or humans tripping on raised rails. The Sixth Avenue streetcar line, the first major addition to the streetcar industry since the New York and Harlem Railroad two decades earlier, laid down 3.75 miles of girded rail in 1852. A financial success, the Sixth Avenue streetcar line spurred on new lines throughout Manhattan and Brooklyn. The Second Avenue streetcar line followed suit that same year (1852) as did the Third Avenue line, and then the following year, the Ninth Avenue line. The Central Park, Forty-Second Street, Eighth Avenue, Bleecker Street, Twenty-Third Street, Houston/West, and other streetcar lines, including the Brooklyn City line, were not far behind. With the city crisscrossed with streetcar lines, rules were put into place banning steam-powered railroads (including the New York and Harlem Railroad) from operating below Forty-Second Street, which was henceforth officially streetcar country.

Early on there was some pushback against streetcars from store owners, who owned the street frontage associated with their shops. Such owners, often called abutters or fronters because they paid for repairs in the street space associated with their shops, feared the dynamics of life on the streets would change for the worse. The store owners worried in particular that streetcars would encourage through traffic, reducing the numbers of people congregating to shop, among other activities. In time, though, as it became clear that streetcars were increasing rather than decreasing revenues, the store owners' resistance faded away.

For the most part, streetcar lines were quite profitable. Costs for the horses, the stables housing them, and the streetcars (which usually cost about $1,000 each and could last fifteen years) were more than compensated for by revenue from fares. In 1857 alone, the Third Avenue streetcar line pulled in $405,278 in fares, the Eighth Avenue line $341,471, and the Sixth Avenue line $267,485. By 1860, New York City streetcars carried somewhere between 35–45 million passengers annually and by 1870, that number had skyrocketed to 115 million passengers a year. When Cornelius Vanderbilt's Grand Central Station, modeled to some degree on London's St. Pancreas, opened on Forty-Second Street in 1871, it acted as a hub connecting three steam rails—the Hudson, New Haven, and Harlem Railroads—which brought in upstaters, and their money, with the many streetcars now traveling south from northern New York State to New York City (figure 1.4).

The new streetcars showed marked improvements from those originally designed by Stephenson for the New York and Harlem Railroad. There were now two basic models: a "bobtail"—that to

**Figure 1.4**
The Third Avenue streetcar depot, 1867. From New York Public Library Digital Collections, accessed November 24, 2025, https://digitalcollections.nypl.org/items/3654f180-c559-012f-8581-58d385a7bc34?canvasIndex=0.

some looked like the feathers on a bobtail bird, though it is not clear why—and a double-ender. Cars typically ran about sixteen feet long, weighed a ton and could seat twenty or so passengers. Both bobtails and double-enders had a single compartment, rather than the three compartments as in older models. Passengers boarded from the rear—rather than in a stagecoach-like manner by side doors—and sat on seats that ran the length of the car and were parallel to one another. Windows lined both sides of the car and many had ornate paintings below them with the name of the line above, written in gold or silver leafing. Models running during the summer were often open on the sides to help with scorching temperatures. The top of the streetcar sometimes had a clerestory—a raised section with windows—and some cars came with a ventilator which circulated fresh air. Streetcar lines used oil lamps to illuminate cars at night and ventilators on the clerestory were especially welcomed on windowed streetcars running those evening routes.

Bobtails were the smaller model of the two, often powered by one or two horses. These streetcars were single ended, meaning one end was the front and one the back: the car needed to be turned around—either by the driver or on a swivel that Stephenson patented—to head back on its return route. Double-enders were larger than bobtails, and had a place for a driver's seat and a place to hitch horses at both ends of the car, and so when a driver reached the end of the line, he and the conductor (if there was one on board) could move to the other side of a car, unhitch the horses, walk them to the opposite end of the car, and head directly to their return route.

"The convenience afforded by Street Railways," wrote New York City Mayor John Hoffman in 1866, "tends pretty equally to all classes of our citizens." Which classes and when largely depended on the time of day and what part of the city you were in. On the west side, streetcars were more likely to be "patronized by the rich," while crosstown, "the easterly lines accommodate . . . many operatives, tradesmen, clerks . . . who work downtown, but [who] prefer the wealthy location of suburban residences to the filthy noise of the tenement houses in the lower part of the city."

Until two years prior to Mayor Hoffman's remarks, "pretty equally to all classes of our citizens" did not apply to African Americans in the city. The Sixth and Eighth Avenue lines had "colored cars," with company directors making a separate but equal argument: African Americans rode in them, or they didn't ride at all. A number of earlier court cases had challenged segregated streetcars, but had failed. Then, in 1864, Ellen Anderson, an African American woman whose husband died fighting for the Union in the Civil War, had the temerity to board an Eighth Avenue streetcar that was not labeled "colored." "The conductor came up and said I could not ride in it," Anderson told the Police Commission when she brought her case there. "I told him I was sick and wished to go home . . . he started up and put his hands on me; I told him not to touch me: then he went out and brought in a policeman. . . . I got hold of the straps of the car and then they both pulled me and dragged me so that I was very sore and they tore my hand and at last succeeded in dragging me into the street." This time, justice was served, as the Police Commission ruled that "there was no law against [African

American] people riding in the cars, and they [the Eighth Avenue line] had no right to make such a law."

Other metropolises both in the United States and Canada followed New York City's lead on adopting streetcars. In 1861, the city of Toronto (that in a little more than a decade would be the first urban center hit by the Great Epizootic) had granted Alexander Easton's Toronto Street Railway a thirty-year franchise. "The time was set for the starting of the first horsecar," the Toronto *Globe* reported on September 10 of that year. Large crowds had gathered, "and all were bent on enjoying themselves. . . . When the first car came out of the depot and was placed on the track, a grand cheer arose from the assembled multitude. . . . The National Anthem and three cheers for the Queen welcomed the appearance of the cars." Things did not go quite as smoothly as hoped. "After proceeding a few hundred feet to Bloor Street, the first car ran off the tracks, but several passengers soon had it back in place. . . . The passengers treated the delay as a joke, and the crowd [was] always ready to give a shove or a lift to keep moving."

In time, the Toronto Street Railway was running routes on Yonge Street, King Street, Queen Street and more, carrying thousands of commuters each a day. Tickets cost five cents and the charter that Easton had been granted required his streetcars not to exceed a speed of six miles per hour and to operate sixteen hours a day, except for during the winter, when they need only be running routes fourteen hours a day.[5]

As they had with omnibuses, Boston, Chicago, and other cities kept a keen eye on the streetcars in New York City. Although Boston's rather eccentric city design contrasted with New York City's systematic grid of streets running east and west and avenues running north to south, both worked well for streetcar routes. The first Boston streetcar route went into operation in March 1856, and included a car purchased from a Brooklyn streetcar company with a "Greenlawn Cemetery" sign still painted on it. It ran between Cambridge and Charles Street in Boston and reported over a million paying passengers in its first year. Not long after, a second streetcar company was running a route between Harvard Square and Bowdoin Square. Fares ranged from 5 cents to half a bit, depending on the company and the specific route, and strips of tickets could be bought at discounted rates. Other competing streetcar companies soon followed suit, including the Highland Street line, the South Boston line, the Middlesex line, and the largest of them all, the Metropolitan line. The Metropolitan line, whose rails ran for twelve miles and covered fifteen separate routes, including the Brookline, Dorchester, Warren Street, Jamaica Hills, and the Mount Bowdoin routes, carried eight million passengers in its first year of operation, all of them transported around Boston by horse-powered vehicles.

By 1860, all totaled, Boston streetcars carried about fourteen million passengers a year (for comparison, the figure that year for a much bigger New York City was forty-five million passengers). By 1870, in a city of 250,000, Boston streetcar companies employed more than five thousand people, and the city was home to sixty-two carriage making firms, sixty-one companies that made harnesses, twenty-nine hay dealers, and fifteen wheelwrights.[6]

Chicago's streetcars began running in 1859, which was also the year that John Charles Haines was re-elected as the city's mayor. Haines's inaugural speech on March 16 was chock-full of references to this new mode of transport. After praising the steam rails for intercity travel and reminding the citizens of Chicago that their city "was a great [steam] railroad centre," he turned to the new intracity streetcar railways. Because ticket prices were typically just five cents, streetcars, Haines told the audience gathered at City Hall, "give the poor man a home in the outskirts of the city, at the same time that they bring him near to the place in which he earns the means to support his family." It was still early on in the development of these streetcar lines, but the mayor told the crowd that he "ardently desire[d] that the benefits of these roads will be extended to all quarters of our city." That ardent desire was indeed fulfilled: by 1863, the southside of Chicago was serviced by the City Railway line, and the western and northern parts of the city by the West Division and North Chicago lines, respectively.

With the success of streetcars, direct comparisons soon were being made between them and omnibuses. In his 1859 book *A Practical Treatise on Street or Horse-Power Railways,* Alexander Easton, who "having received numerous enquiries from various parts of the United States and Europe, in relation to the construction and general operation of Street Railways," asked his readers to imagine: "A wet day,—every corner of the sidewalk crowded with impatient pedestrians, each one anxiously peering up or down the street in search of the particular omnibus among the fifteen or twenty approaching, to carry him home, which with as many more coming in the opposite direction, so effectually choke up the street, that the drays and carts unable to cross at the intersections, render the highway impassable to private vehicles, . . . the omnibuses crowded to excess, cannot accommodate the vexed crowd on the side-walk." Then, Easton painted a very different picture: "Not an omnibus is seen in the whole length of the street . . . drays and carts move with comparative ease, little strips of iron are laid along the street, upon and across which, vehicles pass without inconvenience, . . . there is no crowd, for the little [street]cars glide along rapidly and frequently, accommodating everybody."

Americans were not the only ones making such side-by-side comparisons. The British had dabbled with streetcars on and off, but through the 1870s, intracity transport was still largely powered by omnibuses: in London alone, the General Omnibus Company transported 42.7 million passengers on 592 carriages powered by the company's 6,790 horses. Given the growth of the streetcar industry in America, British railway man John Noble thought this reliance on omnibuses might be a mistake, and he and his colleagues sent a series of surveys to the mayors of "the principle cities of the United States and Canada," and published their results, which were presented to Parliament in 1866 in a fifty-six-page report entitled *Facts Respecting Street Railways: The Substance of a Series of Official Reports from the Cities of New York, Brooklyn, Boston, Philadelphia, Baltimore, Providence, Newark, Chicago, Quebec, Montreal, and Toronto.*

The Noble Report had a particular emphasis on "the effect of Street Railways upon ordinary traffic . . . the convenience of the carriages compared with omnibuses [and] the safety of the system as regards both pedestrians and passengers." Members of Parliament reading the report were barraged with data on the superiority of streetcars over omnibuses.

In the large cities of North America, omnibuses had clearly lost the race to control the streets of metropolises to streetcars. That's not to say that they disappeared from the lives of urbanites—New York City, for example, was still granting hundreds of omnibus licenses every year—but rather that by the early 1870s, before the vast majority of urbanites ever heard the word "epizootic," horse-powered streetcars and omnibuses were the primary force transporting people in major cities across the country. But horses did so much more than that. They connected people with friends and family, brought them to the theater, to musical productions, to new stores, to sporting events, and more. They powered all the services that make a city run. They were the energy behind a construction industry on the cusp of erecting the Gilded Age residences and buildings that we admire to this day. Horses moved food, drink, dry goods—virtually all the goods that stocked the shelves in everything from mom-and-pop stores to the new "department stores" that were starting to dot major cities.

Such reliance on horses as the beating pulse of urban centers comes with a price. When such a system breaks, life can come to a standstill. As the *Nation*'s 1872 editorial warned its readers, we treat horses as machines, and not especially valuable ones, particularly in large urban settings where: "we take no precautions against the overcrowding, filth, and want of light and ventilation of city stables." Just the sort of environment where an epidemic might catch fire.[7]

# 2
# Vignettes

On the outskirts of Toronto in September 1872, the horse flu that was dubbed the Great Epizootic began what at the time was an unparalleled march across virtually all of North America. By the time the epidemic had burned out in the summer of 1873, it had also landed in Guatemala, El Salvador, Santa Domingo (today's Dominican Republic), and Cuba, though the damage in those places was minimal compared to North America. There were even rumors that the Great Epizootic had made its way across the Atlantic to England and Ireland.

In North America, the list of cities and towns hit by the Great Epizootic runs pages long. Toronto, Montreal, Quebec City, St. Johns, and more in Canada. In the United States, New York City, Chicago, Boston, Philadelphia, Detroit, Hartford, Pittsburgh, Cincinnati, Cleveland, Indianapolis, Louisville, St. Louis, Nashville, Baltimore, Washington, DC, Charleston, Little Rock, Atlanta, Richmond, New Orleans, Houston, Omaha, and San Francisco all fell prey to the epidemic, as did countless smaller cities and towns. Indeed, records show that the Great Epizootic visited and wreaked havoc (to different extents) in every state and territory except the

Oklahoma Territory (where there were no recorded cases of infected horses, though there may have been unrecorded cases).

We will march along with the Great Epizootic, step-by-step, from its inception to when it burned out; but for the moment, four vignettes—from Canada's capital city of Ottawa, New York City, Boston, and Chicago—will provide a sense of what is to come.

About a week and a half after cases of horse flu were reported in Toronto, the city of Ottawa, approximately 250 miles northeast (and sitting at the border between the provinces of Ontario and Quebec) was hit. The disease moved from the Toronto area up to Ottawa along at least two routes. One was a stepping-stone path, where local movement between cities and towns by horse-powered vehicles allowed a slow spread of the disease north. A much faster, therefore more epidemiologically dangerous, route was by steam railroad, whose cars were often filled with many horses being transported long distances.

Ottawa had been the country's capital for more than a decade and had its fair share of steam railways. In the 1870s, people sent horses by railroad from town to town and city to city to fill the seemingly never-slowing need for more of the animals that powered their lives. Horses as well as cattle, pigs, and other domesticated animals rode in what were called standard boxcars or stockcars that were built with animal transport in mind: the Erie Railroad, which operated around the United States-Canada border, owned 600 stockcars and more than three thousand boxcars (though what proportion of these cars may have transported animals is not known) and another railway reported shipping more than one hundred thirty thousand (live) cattle and horses in the year prior to the epidemic.

The most likely steam rail route by which horse flu moved to Ottawa would have been the Grand Trunk Railway. Trains, some with boxcars and stockcars, departed Toronto often, heading northeast through Oshawa and Kingston to Brockville. From Brockville, the Brockville and Ottawa Railway line would have moved horses due north to Smith Falls and then Almonte, with a final leg east from Almonte to Ottawa. While railway speed depended on many factors, including terrain, rail gauge, weight of cars, weight of cargo, engine strength, and more, based on the available data, twenty miles per hour is a reasonable estimate of the speed of trains between the Toronto area and Ottawa.

Building in time for stops along the way and for the switch of railroads, the trip from Toronto to Ottawa would have likely taken about two days. Any horse that boarded in Toronto and was infected and incubating horse flu might only have just started to display symptoms, if even that, when arriving in Ottawa. Along the way, an infected but asymptomatic animal could potentially infect all the other horses in its boxcar or stockcar.

In the fall of 1872, Ottawa was home to 6,500 adult horses and more than 1,700 colts and ponies that powered thousands of "vehicles for transportation." Lumber was an important component of the Ottawa economy and horses played a critical role in hauling timber from place to place. They also powered many other components of the economy. And starting just a few years before the Great Epizootic struck, for a fare of six cents paid to their driver, they also pulled the streetcars of the Ottawa City Passenger Railway—cars built by the John Stephenson Company that had developed omnibuses and streetcars in New York City. When there was a question of whether the government would support Ottawa City Passenger Railway's request for a charter, the *Ottawa Citizen* turned to provincial pride to rally the citizens of the city: "We are about to become the Metropolis of British North America. The Capitals of Nova Scotia and New Brunswick, of Upper and Lower Canada, all four have street railways," the paper noted, so "shall not Ottawa, the Capital of all be as well furnished?"

As trains arrived in early October, Ottawa was a horse flu time bomb waiting to explode. The first sparks were recorded on October

12, and so the disease likely landed in the city a few days prior. A week later, on October 19, the *Ottawa Citizen* reported that "all the horses of the Street Railway Co., but six, are sick." The Ottawa City Passenger Railway claimed that its horses were sick not because of "the prevailing epidemic, but from the unprecedented severity of the weather and the bad state of the streets," though that seems somewhat suspect, and may have been an attempt to avoid panic among its passenger base, as just two days later the same newspaper reported that "the epizootic amongst the horses in this city has become more serious than was at first anticipated, and has assumed such a violent form as to cause considerable anxiety to horse owners." That article went on to claim that the disease had "inserted itself into nearly every public stable in the city and a large number of animals owned by private citizens are suffering from the malady. . . . It is computed by an experienced veterinary surgeon that there are not 50 horses in the city free from the disease."

The city's leading businessmen soon weighed in on the economic costs of the horse flu that was slowing commerce to a drip. They estimated that a large market on Washington Street was losing $50,000 a day as the result of the epidemic, in part because farmers couldn't find enough horses healthy enough to deliver produce into the city. Newspapers were reporting that the carcasses of horses that succumbed to the disease were being haphazardly "carted to the woods . . . at the west end of the city. If they are not buried before the snow falls, they will render that locality uninhabitable next spring and may possibly breed a pestilence."

Panicked businesses and individuals were being swamped by advertisements for cures. "All these horses at present attacked could be readily and speedily cured by the Turkish bath, without any medicine or suffering by the poor animals," read one such ad in the *Ottawa Citizen*. To make matters even more chaotic, new names were being created for the disease—hypozymosis, epidemic catarrh, equine epidemic, and horse plague to name just a few.[1]

Less than a week after the *Ottawa Citizen* reported that virtually all of the horses that powered the city railroads were sick, the Great Epizootic struck in New York City, Boston, and Chicago. The disease and its crippling effects on life in these metropolises was front-page news, day after day, for months. On October 24, 1872, for example (though one could find something similar on any day from late October through December), major stories on the Great Epizootic appeared in the *New York Herald, Boston Globe,* and *Chicago Evening Post.* The *New York Herald* headline that day read "POOR BEASTS," with subheadlines decrying "AN ALARMING PESTILENCE" and "Manhattan Island Turned into a Vast Horse Hospital." In a huge story that covered all six columns on one page, the *Herald* story opened with "the pestilence among the horses is rapidly growing in the strength of its destruction and there are now not less than 15,000 cases requiring medical treatment. . . . The disease appeared yesterday in its most hideous form as the streets were crowded with infected horses." And this was not simply a matter of concern for those who employed horses for the endless tasks that powered New York City: instead, the *Herald* told readers what they no doubt already knew—that "throughout the city everybody was discussing the horse disease."

The *Herald* sent a reporter to visit the Bleecker Street line, one of the many that employed horse-powered streetcars to transport the citizens of New York City to and fro. When he asked one of the drivers about the conditions of the hundreds of horses in the company's stable, and whether there were any very sick horses, the driver replied, "Piles of 'em. They ought to be all very well wrapped up with red flannel and put to bed. Every horse on this line ought to have a bucket of hot rum with half a pound of butter poured into him and then let him rip and perspire an hour or two." Apparently happy to be interviewed by a newspaper, the driver wasn't through. "All our horses have got the chills," he added. Similar reports, the *Herald* told its readers, were coming in from many other streetcar lines and a suite of liveries and stables across Manhattan and Brooklyn.

The *Herald* warned readers there was little reason to think that things would get better in the immediate future. The public clamored for the service of their equine employees, and so people needed "to accommodate themselves to the sight of poor sick brutes that are eyesores in the streets, harnessed up and driven their daily journey . . . an animal in the most advanced state of disease with a revolting discharge of greenish fluid from its nostrils." That said, there was a glimmer of hope, readers learned, as a "veterinary surgeon of the Board of Health is now examining the disease in the city and will make a report upon the subject to the board. . . . A very important and interesting report is expected."

The *Boston Globe* article from that same day, October 24, 1872, was shorter than the *New York Herald*'s, but no less dire in tone. Titled "The Disease in Boston," it was in fact about two diseases: "one of which affects man and the other his most trustworthy and useful friend the horse." Smallpox had broken out, the paper noted, and much was known of this disease, "but [of] the latter usually called the 'horse disease,' little is generally known, as it has but lately made its appearance in our midst." But what was known was very disconcerting. The *Globe* printed a list of streetcar lines, stables, and other enterprises reporting horses down. The prior day the *Globe* editors had run a story in which they informed Bostonians that due to the seriousness of the situation they would provide the advice of a Dr. Elliott, who recommended that "the stable be well ventilated, the horses blanketed, and chloride of lime sprinkled through the stable every morning." In addition, the doctor instructed that "the nostrils should be sponged out two or three times a day if the mucus adheres thereto. The food should consist of bran with a little oats in a moderate quantity of hay." As for a prescription per se, numerous were provided, including linseed oil mixed with turpentine and a mixture of nitrate potash, antimony, and digitalis. Though the paper wouldn't vouch for the effectiveness of such prescriptions, the story ends by reporting that one prescription had "been tried and a gentleman who had 150 horses saved them all."

Compared with New York City and Boston, the disease was a bit slower to reach more westerly Chicago. But soon enough it appeared. On October 24, 1872, the *Chicago Evening Post* struck fear into the hearts of readers in the very first sentence of the article "HARROWING TO THE HORSE": "The pestilential breath of that mysterious disease which has been . . . decimating the equine population of Canada . . . has been born[e] to us on the oriental gales: and owners of horses are already being seized with embryonic panic." The tone got no less ominous as the article went on: "Our horses are attacked. The disease has made its appearance in some of our city railway barns and will doubtless, unless some great preventative can be discovered, spread with alarming rapidity and perhaps fatal effect. . . . It spreads with wonderful velocity and clears out a stable in remarkably quick time."

The disease, the *Evening Post* explained to readers, was fairly simple to diagnose: "The afflicted beast first looks dull and drowsy and does not manifest its usual degree of appetite. Then the eyes begin to run water. The nose follows the example . . . the gland becomes sensitive and inflamed . . . a hacking cough generally accompanies the disease." Sometimes things got much worse as, when "hemorrhages of the lungs . . . [and] rheumatic inflammations" appeared. Even if horses recovered, it was not a pleasant recovery. "The degree of prostration," the *Evening Post* wrote, "which follows even the lighter phases of the affliction is so great that the equine sufferer is generally *hors du combat* [out of action because of injury in combat] for a month or five weeks."[2]

By late October 1872, Ottawa, New York, Boston, and Chicago were in panic mode. And that was just the tip of the iceberg. To understand the evolution of that iceberg, we need to start at the beginning: in the sleepy town of Markham, Canada.

# 3
# Outbreak

Markham, twenty miles due north from Toronto, was like so many other small townships in Canada's Ontario Province. First settled in 1790, it sits between the Rouge and Don Rivers: an early land auction made particular note that the latter river "affords an excellent salmon fishery, large enough to support a number of families." The soil around Markham was rich and was tilled accordingly, with residents sowing wheat, barley, oats, peas, turnips, corn, and potatoes. Orchards and pastures also speckled the landscape. Mills were built to grind oats and barley, a distillery was in place to produce spirts, and other industries included a foundry, tanneries, a brewery, and a sawmill. The 1871 census pegged the population at 8,152, and a somewhat later report noted the city (and surrounding pastureland) was also home to 2,829 horses.

In late September 1872, the temperature was mild and comfortable, dipping down into the 40s (F°) at night and rising to the 60s at midday. During the last few days of the month, people began noticing that their horses, part and parcel of the town economy, were getting sick with an ailment first described as catarrhal fever. Catarrh refers to inflammation of the mucous membranes, especially those in the nose and throat, and those first sick horses were running a fever and showing signs of inflammation: shortly thereafter, a hacking cough and profuse discharges of yellowish-white mucus from their nostrils were heard and seen. Along with these symptoms, sick horses had limbs that were cold to the touch and were showing a telltale sign of an equine in distress: a "staring coat," where fur stands up with hairs clearly distinguishable rather than lying flat in a lustrous mat. Soon, this ailment would be known by any of more than a dozen names, including horse flu, and ultimately the Great Epizootic.

The original diagnosis of catarrhal fever was issued by Dr. Andrew Smith, Canada's leading veterinarian; soon Smith began calling the disease influenza. Smith had migrated from Scotland, where in 1861, at age twenty-seven, he had graduated with high honors from one of the finest veterinary programs in the world at the University of Edinburgh. With a degree for veterinary surgery in hand, he moved to Canada after being recruited by the Board of Agriculture of Upper Canada to train veterinarians. Five years later, Smith, along with Duncan McNab McEachran, a fellow graduate from the University of Edinburgh who had also been coaxed to Canada, published *The Canadian Horse and His Diseases*. Together they also founded the Upper Canada Veterinary School, which a few years later, after McEachran had moved on to Montreal, became the Ontario Veterinary College.

As with the outbreak in Markham, when the horse flu struck ten miles to the southeast in Scarborough—a town home to about forty-five hundred people and two thousand horses—the records for the first cases are sparse and note only that the disease appeared during the last few days of September, rather than on a specific day: in fact, whether the disease appeared first in Markham or Scarborough can't be determined with certainty. The nature of epidemics is such that the date when the disease is first recorded is almost always later than when the first animals become infected. For one thing, in the case of horse flu, people may simply not notice the earliest cases and only begin taking note when some nontrivial number of animals show symptoms. For another, infected animals will rarely display symptoms immediately. People, including veterinarians, in 1872 did not yet know that the incubation period for horse flu is between twenty-four and seventy-two hours: during this period, horses can be infected and spread the disease, but show no symptoms. During the incubation period horses would have been employed in their typical services, as there was no apparent reason not to.

The day that the earliest cases of horse flu were recorded in the metropolis of Toronto—seventeen miles due south of Markham and thirteen miles southeast of Scarborough—is known. "On Monday, the 30th of September," Andrew Smith wrote to Adoniram Brown Judson, assistant sanitary inspector for New York City, "I found fourteen horses affected in one stable."[1]

It is next to impossible to know for certain the exact path the horse flu took from Markham and Scarborough to Toronto: indeed, there may have been, and likely were, many paths. The three locations are far enough apart that people would rarely have walked between them, but close enough for people to have often made a trip by horse and buggy, stagecoach, or by saddling up their steed. While people and their horses moved between all three locations, the flow would have been heavier toward Toronto, for all the resources that a large city offered.

For the sake of argument, suppose that horse flu first broke out in Markham and then soon thereafter in Scarborough. Even before it headed south to Toronto, it might have moved back and forth between Markham and Scarborough, with horses from one infecting the other and then vice versa, perhaps numerous times. From there, horses in Toronto might have been infected by visits from their equine peers from Markham, Scarborough, or both. And once the disease was in Toronto, it might have moved back up to either Markham and Scarborough or both, and then back down to Toronto. And that's just a few of the many permutations possible: it could be, for example, all Toronto infections were the result of interactions with only infected Markham horses or with only infected Scarborough horses.

With a population of fifty-six thousand, Toronto was Canada's third-largest city (behind Montreal and Quebec City) and played the part well. Downtown was a vibrant place replete with banks, printing houses, taverns, restaurants, shops, manufacturing plants. and houses of worship, including the stunning St. Michael's Cathedral, with its Gothic Revival architecture, on Bond Street. The Toronto Stock Exchange helped power the city's commerce and the Palace of Industry, a scaled-down version of the famous Crystal Palace in London, was a sight to behold. Sitting on King Street West and measuring 256 feet long, nearly a hundred feet wide and fifty-five feet tall, the Palace of Industry was home to 47,000 square feet of floor space and could host 8,000 people.

The 1871 census lists Toronto as home to 1,875 adult horses. That number is roughly on par with the number of horses in Markham and Scarborough, the difference being in those smaller towns horses were primarily used in agriculture and for personal transportation, while in Toronto they were used in the construction industry, to move goods, to keep stores stocked, and to power the city's streetcars and hacks that took people to work, to shop, to enjoy a dinner and show, and more.

Visitors to Toronto could hop on one of the Toronto Street Railway streetcars or into a hack to find any number of hotels, from the barebone variety to high-end establishments like the Rossin House Hotel at the Corner of King Street and York Street, whose advertising pamphlets boasted of a "splendid commodious hotel . . . finished and furnished with every regard to comfort and luxury . . . the most exceptional first-class hotel in Canada." Along with city residents, hotel guests in Toronto could jump on a streetcar to theaters and entertainment venues galore. They might alight at the new Royal Lyceum staging *Whose Wife* as well as *Dick the Detective,* in which "sensation follows fast upon sensation." Or, perhaps they'd take a ride to soak in O'Brien's *Four Great Shows*, billed as *"The Largest Traveling Exhibition in the Known World,"* or to the Great Golden Menagerie. Another option would be a nice dinner at a restaurant and then boarding a streetcar or a hack to transport them to a performance by German singer and composer Hermine Küchenmeister-Rudersdorf, who had appeared at the Royal Italian Opera in 1855, and was performing in Toronto. When the streetcar horses fell victim to the Great Epizootic, transport to all those locales and more ground to a halt, as did commerce.[2]

Once horse flu struck Toronto, it spread with alarming speed. Horses were getting sick in the stables of the Toronto Street Railway, as well as in Coutt's Stable, Craig and Thompson's Stable, Temple's Stable, Nowlan's Stable, Darragh's Stable, and more. Victims fell ill for anywhere from a few days (on rare occasions) to two to three weeks. "An epidemic," the *Toronto Leader* told its readers on October 5, 1872, "or if not an epidemic, something very like one, appears to be spreading amongst the horses of this city just now." As with most infectious outbreaks, the number of sick horses reported early on varied, depending on the source. On October 7, a week into the epidemic, the *Toronto Mail* reported, "The disease has attacked hundreds of the animals. . . . The epidemic is not confined to any particular stable but has appeared in all the large ones in the city." That same day, the *Hamilton Spectator* reported that veterinarian Andrew Smith "had treated no less than six hundred diseased animals," a number that the *Ottawa Citizen* also cited. "This distressing disease continues unabated," the *Toronto Mail* noted the next day. "Nearly every other horse is affected by it. Coughing and sneezing will be heard continually from the poor animals, and general symptoms seen of a kind of influenza."

While the exact number of infected horses early on will likely never be known with certainty, the six hundred cited as treated by

Dr. Andrew Smith serves as a likely minimal number, though unsupported claims such as the *Canada Farmer's* that during the first two days of the outbreak alone "two thirds of the horses . . . became affected," were also making the rounds. Alarmist and hyperbolic as it was, the *Perth Courier's* October 11, 1872, report that Toronto was a "vast hospital for diseased horses" did describe the *general mood* of the city. In all the hysteria, newspapers were soon bandying about new names for the disease—such as hippodiptheria, which was tame compared to hippolaryngialmorbificality and hyposnoolcalrunnosis—to sound more erudite than their rivals.

If we take those six hundred horses that Andrew Smith doctored early on as a minimal figure and assume that number grew significantly as the epidemic spread and burned hot through mid to late November, then at least one third of Toronto's horses, and likely many more, were down with the flu at some point. All the services those horses provided, from hauling construction material to delivering goods to stores to transporting people to work, friends, and amusements, were slowed to a crawl for more than a month. Businesses suffered as a result: 'The market has been badly supplied with everything all week," the *Kingston Daily News* reported, "[a] cause of which is . . . to be found in the horse disease."[3]

As to what was done to treat sick animals, in a report on the epidemic he published in 1873 after the Great Epizootic, Andrew Smith repeated recommendations and diagnoses—picked up widely by newspapers—that he made in real time during the epidemic, such as a treatment of "chlorate of potash or iodide of potassium, given in one drachm doses morning and night." If things took a turn for the worse, Smith suggested that "relief may be afforded by a mustard application to the throat and windpipe. The nostrils should be sponged out several times a day with tepid water, and afterwards washed with a weak solution of bromo-chloralum or carbolic acid."

Disinfecting the stables where sick horses would be spending vast amounts of time was also critical and Smith suggested the stalls be doused daily with chlorate of lime or iodide of potassium. Smith also reminded those with afflicted animals not to be taken in by snake oil salesmen hawking "complex mixtures which are held out as specifics to cure this and every other disease." He had good cause to be concerned there. Soon after the disease struck the city, Toronto newspapers were full of advertisements for "cures." Many of these provided vague descriptions of the cures being touted, making it hard to know what they even were, as in Thirkell and Perrin's cure for "the horse disease," which simply noted they were providing a "medicine which has proved effectual so far in every case," or Drs. Craig and Thompson's "Remedy for the Horse Distemper," which touted the "wonderfully efficacious results," they had obtained, leading them to assure "the public that no medicine can be obtained more speedy and successful than these powders. One trial will convince." There were others, like Thomas Thompson & Son, trying to cash in by pitching nonmedicinal aids. In their October 14 advertisement in the *Globe*, they asked readers, employing a very early usage of the term *epizootic*, "is your horse sick?" For if it was, they urged, "try our non-epizootic blankets," presumably meaning blankets that had never been on horses with the flu, though if the public thought the blankets themselves had anti-epizootic properties, so much the better.

In assessing the cause of this epizootic influenza, Smith thought it was "probably depending upon some atmospheric influence, as is shown by its appearing simultaneously over a large extent of country, attacking all kinds of horses, young and old, in good condition and poor," which is very much in line with what he had written about horse flu in his 1867 book with fellow veterinarian McEachran, *The Canadian Horse and His Diseases*. That assessment might strike our modern ears as strange. To make sense of it, we need to understand that in late 1872, as the Great Epizootic was kicking into gear, the germ theory of disease—the idea that microbes, invisible to the naked eye, can cause disease—was still only in its infancy. Louis Pasteur's work demonstrating that bacteria can cause disease, as well as Joseph Lister's studies of antiseptics, were known by physicians and some veterinarians, as was John Snow's work demonstrating that a waterborne microbe (that was later identified as *Vibrio cholerae*) was the putative cause for the 1854 outbreak of cholera at the Broad Street pump in London, but germ theory was by no means widely accepted when the Great Epizootic struck.

The "atmospheric condition" that Smith took to be the cause of the epidemic stems from his adopting the main alternative to germ theory at the time. This idea, known as the miasma theory of disease, can be traced back as far as the Greek physician Galen and the work of Hippocrates. Miasma's etymology is from the Greek verb μιασμα, which means to soil or to corrupt. In a nutshell, the miasma theory of disease argues that inorganic particles in the environment cause disease.

The miasma and germ theories each make claims both about whether disease-causing agents are alive (biotic) or not (abiotic) and whether these agents are transmitted, directly or indirectly, from host to host; that is, whether they are contagious or not. Germ theory holds that many diseases are caused by microorganisms, while miasma theory posits that abiotic particles—carried in noxious chemicals arising as vapors from standing bodies of water or from the "bad air" that results from rotting, fermenting organic matter—often cause disease.

Germ theory supposes that the microorganisms that cause disease are often spread from host to host, either directly, via host-to-host contact, or indirectly, for example, when horses drink from the same trough, or, in other cases, by some insect vector that spreads disease. In the eighteenth century, some proponents of the miasma theory, including Jean-Joseph Menuret de Chambaud, who wrote the "MIASME" entry for Diderot and d'Alembert's massive *Encyclopédie*, suggested that particles emitted by bad air and bad water could be transmitted from individual to individual, but the vast consensus by the time of the Great Epizootic was that disease caused by miasma was not contagious, but rather that infected individuals became ill independently, as a result of exposure to a common miasmic agent. One environmental historian has called the nineteenth century "the miasma era," in which "miasma was considered dangerous because it was thought capable of causing the bodies of people who inhaled it to decompose or ferment. . . . Sources of miasma appeared ubiquitous; contemporaries traced it to stagnant marshes, garbage, horse manure [and] burial grounds."

Andrew Smith's position in 1872 was that some sort of abiotic miasmic agent that horses were exposed to had caused the disease to appear and that the miasmic agent was not transmitted from horse to horse. He was vague about what the exact miasmic agent was or where it arose, except that it was "atmospheric." As horse flu began spreading farther and wider, Smith never tackled the question of how bad air or bad water could explain the emergence of the disease in so many different locations. Did the miasma appear independently in Markham, Scarborough, and Toronto, as well as other smaller locations where it was being reported? Did it somehow move, either on its own, or with the unwitting help of humans? Smith doesn't say, though in the past, proponents of the miasma theory were certainly creative when it came to solving such problems. When the bubonic plague hit Venice in the fourteenth century, some suggested that it came with the bad air in crates that were stored on board ships.

Smith's hypothesis that horse flu could be understood as miasmic in origin helps explains his references to the dangers of "impure air," and recommendations such as "[it is of] the greatest importance [to] allow a sufficient supply of pure air . . . well-ventilated stables are found to be of the utmost benefit in mitigating the symptoms, and in hastening the period of convalescence." It is true that even if he had adopted the germ theory of disease, Smith might very well have made such general-purpose recommendations, but his choice of language shows just how powerful the miasma theory still was in the fall of 1872.

The idea that each horse got the flu independently as the result of common exposure to some miasma, rather than by exposure to other sick horses carrying germs, meant that there was no reason for Smith to have suggested quarantining animals, and he issued no such recommendation. The upshot of this was that horse flu could spread in stables, as well as when horses who might be ill, but not noticeably so, were put back to work. But the situation in regard to the spread of the disease was actually even more dire because Smith also recommended that when "animals are not severely affected, a little moderate work does not appear to be hurtful; and in almost all cases it is desirable to give a little walking exercise daily in the open air." However, both working the sick animals and exercising them only fueled the horse flu epidemic, further compromising horses' immune systems and increasing the risk of sick horses potentially infecting healthy ones.

Many newspaper articles reported Smith's diagnosis and recommendations, describing the doctor as an "eminent veterinarian . . . whose great reputation is a guarantee of the soundness of the advice given." But other papers took issue with his miasma hypothesis: "We warn our readers," the *Brantford Weekly Expositor* wrote, "that they cannot be too cautious about allowing healthy animals to come near those suffering with the disease." Other papers thought Smith's advice that "work does not appear to be hurtful" was downright cruel: "Owners of horses, should bear in mind," the *Hamilton Spectator* told its readers, "that it is very reprehensible working horses when they are suffering from the disease," while the *Sun* warned, "If you wish to save the lives of your horses do not force the poor staggering animals to work." These newspaper articles, along with others like them, did some to mitigate the dangerous effect of the miasma diagnosis and the recommendation that light work was fine for infected horses, but nowhere near enough to stop the spread of the horse flu, which was moving from Toronto, Markham, Scarborough (and likely other locations) in every direction.[4]

The Grand Trunk Railway Company had express trains running between Toronto and Montreal and on October 8, 1872, the first recorded cases of horse flu among Montreal's 3,500 horses were logged by Andrew Smith's colleague, Duncan McNab McEachran, who had opened a practice there and was widely regarded as one of—if not the—city's preeminent veterinarians. The same economic woes and slowdowns that arose in Toronto and Ottawa were soon manifest in Montreal. Perhaps because it was the largest city yet hit, a new economic piece was added to the epizootic puzzle, when certain unsavory entrepreneurs used the bottleneck on stocking shelves and moving goods to begin to charge what the *Montreal Herald and Daily Commercial Gazette* described as "extortionate prices."

A week after the first cases of horse flu were noted, the *Montreal Herald and Daily Commercial Gazette* published a short piece by Charles Geddes, the president of the Montreal City Passenger Railway Company, in which Geddes let it be known that "in consequence of the prostration of the Company's horses from the effects of the prevailing epidemic, the running on the various lines must necessarily be partially suspended." The lack of streetcars became bad enough that one Montrealer wrote in to the *Montreal Star* lamenting that when, a few years back, several hundred people in the city tried to introduce velocipedes—bicycles—as a means of intracity transport, they were laughed off the streets. "Had [their] plan been adopted," the nostalgic writer continued, "we should now find the bicycle a general favorite, and we should look to it as a possible means of temporarily supplying the dearth of ordinary means of conveyance."

Ten days after he recorded the first cases of the disease in *Montreal*, McEachran wrote an editorial piece for the *Montreal Star*: "That the epidemic among horses is rapidly spreading and even now prevails to such an extent that it is impossible they can all have professional attention," he told readers, "I felt it my duty to the public to offer the following hints as to the management of the disease." He first listed the symptoms of the disease and then made recommendations that lined up with what Andrew Smith had recommended, including that moderate work was fine for sick horses. Soon, McEachran was presenting public lectures on the disease to the Natural History Society, where he called upon the government to give grants "annually for the investigation and experiments to discover the real nature, and the best means of treating and preventing diseases of this nature, which [are] the sources of incalculable loss to the country . . . [including] the interruption of business and the loss of daily earnings."

Horse flu continued its march north and east. Quebec City was hit not long after Montreal, and the disease reached St. John, New Brunswick, 450 miles due east of Montreal, no later than October 26, when that city's *Morning Freeman* published the advice McEachran had provided to the papers in Montreal. Less than a week later, the same paper reported a grim scenario. "At least two thirds of all the horses in the city . . . are sick, and the consequent injury to businesses and inconvenience to householders is very great," it told its readers, who no doubt knew from personal experience that "the streets present quite a desolate appearance . . . grocers cannot even supply their city customers." Two days later it was even worse: "It is difficult to get even the work of the most urgent necessity done, as nearly every horse in the city is now sick," The *Morning Freeman* wrote, continuing its reporting on the ongoing epidemic: "The streets, usually so full of life and animation, are idle and unoccupied, presenting a strange, holiday appearance which makes these dull times positively gloomy."

As the disease moved across eastern Canada, a number of patterns began to emerge. Horse flu spread very quickly from horse to horse—"spreading with the alarming rapidity of the Asiatic Cholera," as one account described it—with many, if not most, of the horses in a town or city falling ill in less than a week from when early cases were reported. Horses were usually sick for two to three weeks, but mortality rates were generally very low. Exact figures are not available from Canada, but based on newspapers and reports compiled much later, only a very small fraction of infected horses died. In the language of modern epidemiology, the disease was highly transmittable, but not especially virulent. Diseases with high transmissibility but low virulence tend to move quickly between locations, but also burn out relatively quickly at any given location, which is just what happened in Toronto, Ottawa, Montreal, St. Johns, Quebec City, and other cities and towns in Canada, where horse flu tended to tear through populations, but then fizzled out in four to six weeks.[5]

Epidemics don't respect international borders, and the Great Epizootic headed southwest into the United States: soon horse flu was moving back-and-forth across the border. The earliest record of cases in Detroit, 250 miles southeast of Toronto, was on October 10 (or perhaps October 11), with the disease likely having arrived by moving west within Canada and then heading south along Lake Saint Clair.

At the time, Detroit was a melting pot population with half of its approximately eighty thousand inhabitants having been born outside the United States. Detroiters had been on the lookout for horse flu. Five days before it arrived, the *Detroit Free Press* alerted readers, in a story with a byline from Toronto, that "there is an epidemic among the horses here of a catarrhal character. Almost all the horses belonging to the streetcar companies and livery stables are affected." Once the horse flu was recorded in Detroit, "coming no one knows how," as the *Free Press* reported, it wreaked the havoc it had in other cities. By October 25, at least a thousand horses were or had been afflicted with the disease.

Horses played an outsized part in powering Detroit's economy. In this manufacturing hub horses moved almost everything that needed to be moved from the docks into and throughout the city. They kept the shelves in fur and clothing stores full of the latest new attire, made sure the cigar shops were stocked, and delivered needed materials to tanneries and soap-making warehouses. All of that activity slowed to a trickle when the Great Epizootic struck. When creameries and breweries needed their products moved, they did so only with the aid of horses—but not during the outbreak of horse flu. Rumor had it that horses in the brewery stables were fond of a pint themselves and would not head out for work until they wet their whistles, but, when suffering from the flu, even that enticement was not enough to draw them out of their stables.

Horses powered the streetcars that made big city life in Detroit run. In 1862, the Detroit Common Council had granted a thirty-year franchise to the Detroit City Railway Company to run streetcars holding up to forty people each. Lines were put into place on Woodward, Jefferson, Gratiot, Grand River, Fort, and Michigan Avenues, but traffic along all these lines all but stopped during the Great Epizootic. As the *Free Press* reported on October 20: "Friday morning all the streetcars, except on the Grand River Avenue routes were hauled off, and many of the draymen and teamsters refused to bring out their horses." They refused in part because many of the animals were sick, but also as a "precautionary measure."

The *Free Press* outlined the classic symptoms of horse flu for its readers and gave them a list of putative remedies, including one made by physicians at a meeting of the Homeopathic College who wrote that "from analogy with similar diseases in the human subject . . . the disease would be successfully treated with aconite and belladonna in the primary or fever stage, and arsenic as soon as the fever emitted and debility occurred." Not long after that, the paper reported that "the call on the drug stores was so great . . . that the stocks of the retail dealers were soon exhausted."[6]

Horse flu also crossed the border at Buffalo, likely after the disease moved south from Toronto to Hamilton along the coast of Lake Ontario, then past Niagara Falls, where cases were recorded on the Canadian side at Fort Eerie on October 10. The International Railway Bridge, the first bridge to link Canada and the United States, was not opened until late 1873, but there was a plethora of ferries always in motion, carrying goods and livestock, including horses, between Fort Erie (Canada) and Buffalo.

The first horse flu cases were recorded in Buffalo, which sat close to the northeast shore of Lake Erie, on or about October 13. The city's population was closing in on 120,000, and like in Detroit, people were keeping a keen eye on the horse flu across the border and had learned a week earlier of the outbreak in Toronto: "Almost all the horses belonging to the streetcar companies and livery stables are affected," the *Buffalo Commercial* reported, following up a few days later that the disease in Toronto "had become epidemic and has resulted in this serious interruption of business." By October 15, the same paper had informed its readers that the disease "has made its appearance on this side of the line and in a very virulent form," using the word "virulent" in the general sense of dangerous. William Sommerville, a veterinarian who wrote an article for the *Buffalo Commercial*, turned to the miasma theory to warn that unless the cold weather had "the effect to purify and cleanse the atmosphere of this epidemic, horse owners may expect to see a serious malady among their stock." No such purification and cleansing occurred, noted the *Buffalo Commercial*, as they mocked the miasma theory. "It must be in the air . . . is the unscientific but general answer to those who are asked to explain the 'epizootic influenza,'" the paper editorialized. "It is certain that the air is full of rumors that are calculated to do as much harm as the disease itself."

Soon horse flu was all over Buffalo. "At the present time competent judges estimate that from one-half to two-thirds of the horses in the city are suffering from the disease and its various stages," wrote the *Buffalo Courier Express* on October 19. "The street railway companies have numbers of sick animals in their stables . . . the merry truckmen seen on the Terrace and at the courthouse park who occupy their idle moments in badgering each other in diverse ways much to the amusement (?) and edification (?) of the innocent involuntary spectators are silent, their time and energies being wholly devoted to the treatment of their sick animals."

Omnibus service ground to a halt. Canal horses were sidelined from their unenviable task of towing small boats on the water as they walked along "tow routes." And yet, even in the face of a clear epidemic, many owners kept working their horses, wearing down their already worn-down immune systems and exposing them to uninfected animals. "We pity the poor beasts which are being worked by their owners at the present time period," the *Buffalo Daily Dispatch and Evening Post* bemoaned to its readers, "as they go moping along with drooping heads and tearful eyes, utterly refusing to assume a lively gait, they remind us strongly of some poor individual who having been afflicted with seasickness comes upon the dock of a steamer . . . to get a breath of fresh air."

The extent of the damage caused by the flu is captured in the headline and multiple subheadlines of the October 22, 1872, issue of the *Buffalo Courier*: "THE CITY ON FOOT" read the headline, followed by "GREAT PREVALENCE OF THE HORSE DISEASE," "OVER TWO-THIRDS OF THE HORSES SICK IN BUFFALO," "THE STREETS DESERTED AND BUSINESSES EMBARRASSED," and then an opening sentence reminding readers that "the extent of mankind's dependence upon equine assistance is being vividly brought to you by the widespread demoralization which exists in Buffalo in consequence of the presence of the Canada horse epidemic disease."

Buffalo was a blue-collar city employing nearly a thousand people in the manufacture of pig iron and thousands more in building and construction. Another thousand workers drew their salaries from the clothing industry, the same number from tobacco and cigar production, and three hundred from the city's thirty-eight breweries. Buffalonians built ships and steam railroads, operated flour mills and bakeries, moved tens of thousands of tons of freight along the Erie Canal, and ran the stores and shops in the city. All these operations required horses to move materials and people and now hardly a healthy one could be found: "There are not enough well horses to carry merchandise through the streets," wrote the *Buffalo Commercial*. "Boats on the Erie Canal are tied up from one end to the other and thousands of tons of produce cannot reach market." So profound was the effect of the disease that on October 29, one week before the presidential election, the *Buffalo Commercial* noted that "public interest in the approaching election is more or less tempered by anxiety in regard to the probable duration of the 'horse disease.'"

What made Buffalo especially dangerous for the spread of the epizootic was all the between-city and between-state traffic than ran through the city. It was the western terminus of the 363-mile-long Erie Canal—hence the canal horses. It was also a railroad hub, with the many railways coming in and out of the city, serving as rolling vectors for horse flu: Cornelius Vanderbilt's New York Central and Hudson River Railroad, which headed in many directions including to Albany and then down to New York City, as well as Vanderbilt's Lake Shore and Michigan Southern Railway, which headed to Chicago, and many small railroads, such as the Buffalo, Bradford and Pittsburgh Railroad, all rolled through Buffalo.[7]

Traveling seventy miles east, and slightly north, both by railroad and the stepping-stone path, horse flu was first recorded in Rochester, New York, on October 17. Four days later, the *Rochester Democrat and Chronicle* reported three hundred horses were ill, including many that powered the city's fire engines and streetcars. Though no horses had died yet, the paper estimated mortality linked to the horse flu in general at 2–3 percent, though it is not clear how they arrived at that figure. Businesses scrambled to keep their shelves stocked, doing whatever they could to power deliveries: "All branches of business are more or less affected by the scarcity of horses," the *Rochester Daily Patriot* told its readers. "The merchants are put to great expense in hiring men and boys to transfer goods."

While many worked their horses that should have been convalescing in the stable, the *Rochester Democrat* wanted their readers to know that our better angels, that Lincoln had referred to his Gettysburg Address, were also on display in one "public spirited boy" who had told reporters, "The horse is a useful animal. He has four legs, but they oughten to be no good when he's got the epizoocy. Some wicked men drive him ennyhow and makes him koff enuff to koff himself away. If I had a klub and was big enuff I would go about and nock those men in their head . . . we can get along without them, but the horse is a useful animal and harder to get and keep alive. I wish these wicked mens had the epizoocy instead of the horses."

Horse flu hit Syracuse, eighty miles from Rochester, no later than October 19. Within just a few days, the New York Central Railway, along with more local movement of horse-drawn vehicles, transported the disease from Syracuse to the state capital of Albany, one hundred miles further east, and from there to Newburgh, in the Catskill Mountains, where it arrived about the same time.

The highly transmittable strain of horse flu that reached Newburgh was just seventy-five miles from New York City, the economic capital of the country and home to thirty-eight thousand horses and nearly a million humans. No city in the United States relied more heavily on horses than New York, and so before we examine how everything ground to a halt there when the Great Epizootic struck, it's worth exploring just what that everything was.[8]

# 4
# Horse-Powered

Niblo's Garden Theater at the corner of Broadway and Prince Street had been a fixture in Lower Manhattan since its doors opened in 1837. By the eve of the Great Epizootic, horse-powered streetcars were bringing audiences to Niblo's, and those streetcars were coming in droves, delivering crowds to see performances of *The Black Crook*. A saucy, Faustian-like, four-plus-hour show, with songs like "You Naughty, Naughty Men" and "March of the Amazons" interspersed with eight ballet scenes across its four acts, many, though not all, historians of the theater consider *The Black Crook* to be the first musical performed in the United States.

As streetcars zipped in and out of the area around Broadway and Prince Streets, gentlemen and ladies aplenty, dressed in their finest theater attire, and perhaps having just had an early dinner at African American restauranteur Thomas Downing's Oyster House on Broad Street, or further uptown at Delmonico's on Fourteenth Street and Fifth Avenue, could be seen stepping out of their horse-powered transit and entering Niblo's Garden Theater in the early evening to be part of a *must-see* cultural event. Passing through three sets of walnut doors with stained-glass insets and onto the marble-floored lobby, many would have gazed up at the domed ceilings with their frescoes, and the stunning chandeliers that lit that space, before settling into their maroon, hair-stuffed cushioned seats to be transported to the Harz Mountains of seventeenth-century Germany.

Audience members might have had some sense of what they were in store for when the curtain went up at 7:45 p.m., as advertisements for *The Black Crook* were a staple in New York City newspapers since the musical had opened in 1866. Many would have excitedly awaited the appearance of the poor artist Rondolphe and his love, the village maiden Amina. They would have eagerly anticipated the appearance of the evil Count Wolfenstein and his henchman von Puffengantz and, of course, the hunchbacked master of black magic, Hertzog the Black Crook himself.

When *The Black Crook* curtain fell at around midnight, streetcars with horses at the ready were lined up outside waiting to transport audience members back from seventeenth-century Germany to their homes in late nineteenth-century New York City.[1]

In 1872, New York City was still reeling from the corruption that William "Boss" Tweed, Grand Sachem of the powerful and corrupt Tammany Hall political machine, and his Tammany ring had wreaked havoc upon the city for the prior decade and a half. Despite the chaos that Tweed's ring had caused, New York City survived, which meant that all that makes a city live and breathe, the services that citizens often take for granted, survived as well. Garbage needed to be collected, the dead needed to be taken to their final resting places, streets needed to be cleaned, goods needed to be transported to stores of all kinds, and firefighters and police needed to be at the ready to respond to trouble. All of these services, including the delivery of the mail, one way or another rested on the backs of the horses that powered New York City.

By the eve of the Great Epizootic, the United States Postal Service had come a long way since the Continental Congress appointed Benjamin Franklin the first Postmaster General in 1775. In the

1850s and 1860s, mailboxes began to appear in cities like New York, so that citizens could drop off mail without spending time in line in a post office. But postage paid only for the movement from post office to post office, which meant that one still had to go to the post office to receive mail. Then in 1862, Postmaster General Montgomery Blair proposed all mail be delivered to customers by a new army of sorts—mail carriers paid by the federal government. The idea was to start by stationing that new army in large cities (like New York) that generated sufficient revenues to cover the cost of this new service. Congress agreed and made it law on July 1, 1863.

By 1872, an intricate cross-country system for delivering mail near and far was in place: nearly thirty thousand post offices, from large to small, were spread across the country and on the order of two million pieces of mail were sent each year. In such a complex transportation web, mail moved by many means, largely determined by distance. At a macroscale, steam locomotives moved mail very long distances and at a microscale, in some locations, postal workers hand-delivered mail to customers, but at every other level—the movement of the mail from railway stations to post offices, from mailbox to post office, and from one post office in the city to another—depended on horse-powered vehicles. Of all the links in that complex web, that of the horse was arguably the most important. Without trains, mail could still move within cities, and indeed most mail was intracity. And even sans mail carriers people could still gather their mail at the post office, as they had before: but remove horse transportation and the whole system might fall apart. A similar argument could be made for many of the freight and express companies that were also key to the transportation network because the U.S. Postal Service did not ship packages at that time. These companies did have deep ties to railway companies, but many, like Wells Fargo & Co., the United States Express Company, the American Merchant Company, the National Express Company, the Adams Express Company, and the Transfer Company, all in New

York City, had stables full of horses to move their freight within the city (and to nearby locations).[2]

Like the mail, corpses too needed to be moved. In New York City, with a population near one million, rates of mortality ran high. With a crude death rate of about thirty per thousand per year, approximately thirty thousand citizens of the city would have died in 1872. Those bodies needed to be moved—numerous times, including from the home of the deceased to the mortician, who prepared the body for burial, and from the mortician to the cemetery. A simple horse-powered wagon or cart could be used for the trip to the mortician, while horse-drawn hearses were often employed for the trip to the cemetery. Depending on the social class of the deceased, hearses ranged from simple to elegant and ornate, the latter made of fine wood with windows, sometimes draped with curtains along their sides, and on rare occasion all four sides of the hearse were made of glass. No horses meant no use of hearses. But hearse-pulling wasn't the only role for horses. Caskets sometimes needed to be moved from the mortician to the home of the deceased, again by horsepower. And the very wood that was used to make caskets would have been brought to them by wagons and carts pulled by horses.

The New York City Fire Department was powered by horses and the residents of New York City knew all too well how fire could destroy a city overnight. New Yorkers had heard tales of the Great Fire of 1776, which began just a few months after the Declaration of Independence was signed, but there was no need to reach back in time that far for a devastating conflagration. At 9:00 p.m. on the frigid evening of December 16, 1835, a fire broke out at the Comstock and Andrew's Store on Merchant Street. Within minutes, fifteen nearby buildings were in flames. After depleting nearby water sources, fire crews sent a team to break through the frozen East River for help to douse the inferno. Business owners scurried to their warehouses in the vicinity and pulled out everything they could before their inventory turned to ash. Then "a gust of flame, like a streak of lightning," swept across the landscape, the *New York Sun* reported, "and set fire to the entire mass, which it in a few moments consumed to cinders, and then communicated to the houses opposite." By early morning of the next day the fire was under control. But in the course ten or so hours, 674 buildings in the heart of Manhattan's business district were gone. Ten years later, another conflagration engulfed Lower Manhattan. That one began at 2:30 a.m. on July 19, 1845, at the Oil Merchant and Stearin Candle Manufacturer at 34 New Street. By noon the next day, more than three hundred buildings had burned to the ground.

Despite having suffered three major and countless smaller fires, firefighters were volunteers and not under direct supervision of city governance. It wasn't until 1865, when Engine Company Number 1 went into service on July 31 at Four Centre Street, that the city created the Metropolitan Fire Department (MFD). By 1870 the MFD had morphed into the New York City Fire Department (which eventually became the Fire Department of New York City). Under Commissioner James Hennessey, nine battalions were in place in Manhattan. Horse-drawn fire trucks, often loaded with close to a thousand gallons of water and weighing more than eight thousand pounds, required three or more steeds to power them, while hook and ladder carts carrying seventy-three-foot ladders, were pulled by a pair of horses.

William Burns, Superintendent of Fire Horses for the New York City Fire Department, had his hands full. For one thing, horses needed to be trained that upon hearing a fire bell, they were to leave their stable and move to the fire engine to be quickly harnessed. That training took time, but horses are more than smart enough to learn the task. More difficult was teaching them to stand still once they got to the site of the fire. Like many animals, horses have an innate fear of fire, and only some could be taught to ignore that fear.

Given the extent to which a firefighter relied on his horses, it is hardly surprising that a strong bond formed between the two. "Great is the fire-horse and mighty is his master's love and respect for him," wrote Alfred Downes in his *Fire Fighters and Their Pets*. "The horses are the firemen's pride, their chums and their leisure hours and their partners in danger . . . any needless injury or neglect, any annoyance or teasing, or even a suggestion of brutality to the noblest animal is regarded as a crime worthy of the severest punishment."

Horses did not just power fire engines and hooks and ladders. When people were hurt in a fire, it was horse-drawn ambulances that shuttled them to the hospital. Indeed, the New York City Fire Department coordinated with Bellevue Hospital, which had created the city's first ambulance corps in 1869. By 1870, there were seven horse-powered ambulances, each weighing about eight hundred pounds, and each equipped with a bed, surgical lamp, pillow and blankets, as well as "a box beneath the driver's seat, containing a quart flask of brandy, two tourniquets, a half-dozen bandages, a half-dozen small sponges, some splint material, pieces of old blankets for padding, strips of various lengths with buckles, and a two-ounce vial of persulphate of iron." That year alone, ambulance drivers and their horses responded to 1,401 calls for assistance: some proportion, though exactly what proportion is not known, were in response to fires, which speaks to just how central the horse was to vital city services (figure 4.1).

The New York City Police Department relied on horses as well, though not to the same extent as the Fire Department. Many officers walked their beats, but a small, mounted police unit was already in place in 1872, before the Great Epizootic struck. The unit had been formed slightly more than a decade earlier, primarily to handle both traffic and crowd control at large outdoor gatherings like parades, protests, and New Year's Eve celebrations. Horses by nature are skittish and quick to run when frightened, as might happen for

**Figure 4.1**
Horse-drawn ambulance for Bellevue Hospital. From the National EMS Museum.

any number of reasons on the hectic streets of New York. That was a problem when they were harnessed to vehicles like omnibuses, streetcars, fire trucks, hearses, or postal carts, but it was a seriously dangerous problem when the animal was *mounted*. If a horse reared, bucked, or took off on an uncontrolled run, it could potentially kill an officer (and pedestrians as well). To minimize the chances of that, only geldings (castrated males) were selected for the mounted patrol unit, as they were more docile than mares or stallions.

Training was intense before a police horse was put on the streets. It needed to be able to "respond at once to any tug on the bit, to pressure from the reins or from the rider's legs. It must be ready at call to walk, stop, sidle, back, trot, canter, gallop, or run, and must know how to traverse, or change its lead while galloping," read a *Harper's Weekly* article titled "Making Policemen of Horses." "Then it must know how to stand alone without its rider, and how to 'handle' crowds by pushing or backing into them without harming them." Other training techniques included throwing firecrackers (as young boys were prone to do) to teach horses to remain calm in the face of what otherwise might make them bolt away. In the end, many did not make the cut, but those that did gave the citizens of

New York City yet another reason to appreciate the role of the horse in making their city what it was.[3]

It took *a lot* of horses to run all the streetcars, omnibuses, hacks, mail carts, express companies, fire trucks, hearses, and so on, of New York City. Fortunately, we know how many: the *Annual Report of the Department of Health of the City of New York* for the period of 1870/1871 tallied an exact count of 38,372 horses in city stables, and so a similar number, give or take, were likely present on the eve of the Great Epizootic in the fall of 1872.

When streetcar and omnibus companies, or for that matter anyone, needed additional or fresh horses, there were horse markets aplenty to turn to. Horse market rows—including one that ran the length of Twenty-Fourth Street from Second Avenue to Lexington Avenue—had a reputation for the finest steeds, where "men of horse affinities congregate for the sale, purchase and hire of those 'faithful servants,'" with deals there sometimes finalized at the Bull's Head Hotel and Tavern. There were other horse markets that had a bevy of animals on display every day, like the one which ran between Sixty-Sixth and Sixty-Seventh Streets on Second Avenue.[4]

New York City's thirty-eight thousand horses needed somewhere to live. The city's larger stables were three stories tall, with one floor for horses, one for vehicles, and one for food. In addition to the stables and liveries kept by streetcar companies, omnibus companies, the U.S. Postal Service, the fire department, the police department, the express companies, construction companies, and more, which might employ anywhere from a few dozen to more than a thousand horses, there were scores of smaller stables scattered across the city. In 1870 the Department of Health reported that New York City's Third Sanitary District alone, near the southern tip of Manhattan, was home to 108 stables, and that the Bull's Head Stables, part of the Twenty-Third Street horse market row, had 1,000 stalls.

Smaller stables included the Bradley and Company Stable on Seventeenth Street, Canary's Stable on Mercer Street, Ira Brown's Stable at University Place, and Henderson's Stable on Ninth Street, which housed between a few dozen and a few hundred horses.

Most New York City horses, including those powering streetcars, worked only about five hours a day, traversing from ten to twenty miles, so as to maximize their working lives, which meant they spent the majority of their time in stables. Horses were valuable commodities, and for economic more than humane reasons, much thought went into every aspect of life in stables. Hand-thrashed rye straw was considered the best bedding in a stall, with oat straw, rice straw, pine straw, prairie hay, and Holland peat moss deemed acceptable alternatives. Disinfectants were used, especially in the summer. If horses got sick, veterinarians would be called in. Horseshoes were replaced on a fairly regular schedule in larger stables by an inhouse blacksmith and in smaller stables by a local blacksmith.

Recommendations, sometimes followed, sometimes not, suggested stables should be well ventilated and stalls should be at least nine feet long and four feet wide and the floor pitched at an angle to facilitate drainage. As far as food was concerned, cast-iron food boxes were preferred, and while the method of food preparation—chopped, cooked, steamed, or dry—varied from stable to stable, these boxes were most often stocked with corn, oats, and hay. One street railway stable reported feeding each horse in their stable fourteen pounds of oats and ten pounds of hay a day. For thirty-eight thousand horses, give or take, that amounts to about 190 million pounds of oats and 139 million pounds of hay per year.

All of that food, plus the liquids the animals consumed, produced vast quantities of horse manure and urine. Based on their inspections of all the city stables between April 1870 and April 1871, the Department of Health estimated that a horse passed 30 pounds of manure each day, which translated to 1.14 million pounds of horse manure per day or 416 million pounds per year deposited into the stables and onto the streets of New York City: the horse stable at Central Park alone had a pile of manure that often measured 30,000 cubic feet on its grounds.

*The Citizens' Association's Report upon the Sanitary Condition of the City* complained of the manure heaps on the street: "the stench arising from these accumulations of filth is intolerable." "A FILTHY CITY," a long, five-column story in the *New York Herald* in May 1872, lambasted "the filthy condition of our streets," which are no less than "a disgrace to a civilized community . . . with heaps of . . . horse manure." A few months later, the paper ran another story that told of "a fareful stink . . . with great heap[s] of offal [animal remains] and horse manure" on the streets of New York City. London had a similar, if not worse, problem and employed "crossing sweepers" who carried ladies across manure-coated streets.

In 1870, New York City created the Bureau of Street Cleaning, which that same year claimed to clear almost six hundred miles of dirty streets. But they couldn't keep up with the horses' rate of turning food and water into vast quantities of manure: what they did collect was transported (by horse-powered vehicles) to dump sites near the East and Harlem Rivers, between Thirtieth and Fortieth Streets. As for all the manure accumulating in the stables, it was often carted to nearby outdoor manure pits. Both the stable owners and the city then monetized horse manure—and in the case of the city, "night soil," produced by people and stored in privies that were periodically emptied—by selling it to farmers who used it as fertilizer. For smaller stables, this provided a bit of extra income, but for the larger stables, manure could generate $10,000 a year.

In addition to manure, the stables, and often the streets of New York City, were littered with the bodies of dead horses. Again, both the city and the stable owners, including the large streetcar and omnibus stables, monetized the situation. They would sell the carcasses to animal rendering plants, many of which were clustered around what was dubbed the "offal dock" on West Thirty-Eighth

Street. These rendering plants would use any and every part of a carcass. They sold horse hair to furniture factories to stuff chairs and sofas or to plaster companies to stiffen their product. Hides were sold to leather manufacturers. Candle and soap manufacturers bid on the fat from boiled-down carcasses. Bones were made into knife handles or the lime phosphate in them was sold to manufacturers of matches: even vapors from the boiling-down processes were put on the market to be used in ammonias and insecticides.[5]

In the early 1870s New York City was growing: and growing quickly. At the cusp of the Gilded Age, new buildings, some of them over-the-top ornate, and others more mundane, seemed to rise from the streets on a daily basis. The construction behind all that building was powered, at so many levels, at least in part, by horses. Raw lumber might be rolled into the water to be sent to mills, but finished wood needed to be transported, often by a horse-powered vehicle to a port, if it were going long distances, or to a nearby city or town site, and then to a specific location for use in construction. Horses also played an important role in moving bricks, stone, glass, and other construction materials to a site. Indeed, horses not only moved bricks but were often employed to make them: they powered the pug mills in which clay and water were mixed and then the presses cut the bricks (as well as tile, pottery, and stoneware). At a construction site horses were used to dig foundations, using so-called horse-runs in which the animals would haul wheelbarrows full of dirt offsite. When cranes were needed, or more generally, when items needed during construction had to be hoisted or lifted, horses provided much of the power.

Bricks, mortar, wood, steel, and the like fed the construction industry much the same way that meat, dairy products, grains, fruits, and vegetables fed the citizens of New York City. And, as with the construction materials, almost everything on the dinner table of the average New Yorker would have been transported to its intermediate and ultimate destinations, perhaps numerous times, by horses. Foodstuffs from farms and mills would have made their way into city stores in horse-drawn vehicles, and the deliveryman who brought milk, ice, or bread to their residential customers, or beer to the local pub, would have relied on their horses to help them get the job done. And those residences and businesses would have been warmed in the winter by coal and wood that was delivered by a horse-powered vehicle.

By 1872, shops of every sort lined the streets of New York City, hawking textiles, clothing, hats, shoes, crockery, clocks, flowers, newspapers, furniture, meats of every kind, bread, drink, sweets and confections, art, medicine, and more, often hiring young men as "pullers in," akin to barkers at a circus. A flyer titled *A Peep into Catherine Street or the Mysteries of Shopping*, which touts "such a conglomeration of merchandise of every sort and description cannot be found jumbled together in the same space anywhere else," gives a sense of the pandemonium a shopper in New York City was thrust into: "The crowd that is continually passing, jostling, dodging and buffeting is fully answerable to the vast mass of wares and goods cramped into every building and oozing out of every window, crevice, loophole contour and postures of inducement and temptation." A sensory and visual smorgasbord overwhelmed the senses. "To enumerate the splendid streamers of sanguinary, gold-hued verdant flannels, the chequered surfaces of gorgeous tartan shawls," *A Peep into Catherine Street*, explained, as well as the sight and smell of "perspiring hams boiling and dripping from the awning beams," was beyond the power of words.

All of those stores needed to be stocked. And stocks were moved from ports, farms, warehouses, and other locations by horses. What's more, there was now a growing number of larger department stores whose floors and shelves needed vast quantities of everything a buyer might fancy. Lord and Taylor, whose supply chain depended on horses, had outgrown its space on both

Chrystie and Grand Streets in Lower Manhattan, and on November 28, 1870, moved to an extraordinary new five-story home, designed by architect James H. Giles, uptown at the corner of Broadway and Twentieth Street. During its first three days of operation, ten thousand customers swarmed the store. No small fraction of that crowd was there for the thrill of riding on one of the early steam-powered elevators operating in New York City.

R. H. Macy & Co., smaller than Lord and Taylor, was nearby at Sixth Avenue near Fourteenth Street, as was the A.T. Stewart Store, a five-story establishment between Ninth and Tenth Streets on Broadway that employed two thousand workers in a gorgeous marble building, constructed in Italian palazzo style, with Corinthian columns and eighty-four arched windows. Tiffany & Co., which exploded onto the international scene in 1867 at the Paris Exposition, and by 1872 was seen as providing the country's finest silver work, jewelry, and timepieces, was located in Union Square close to Brooks Brothers Clothing and just a few minutes' walk from Lord and Taylor's, Macys, and A.T. Stewart's.

These stores, along with others such as Arnold Constable & Company on Nineteenth Street and Broadway, which made up the "Ladies' Mile," drew in not only women (and men) of means who might want to purchase an oriental rug, the finest new fashion, or a new piece of jewelry, but also women (and men) of *all* economic classes, as the ornate window displays were designed to inspire the new phenomenon of window shopping. Rather than employing pullers in, department stores used their window displays as lures. In those displays, as an advertisement in the *New York Times* described, one might see "handsome shawls, rich scarves and fine furs, carriage robes" as well as "choice laces, including point applique, round point [and] valenciennes." A trip to the Ladies' Mile in New York was only partly about shopping. It was a leisure event, and, for all those who didn't live in walking distance, participating meant jumping on a streetcar or omnibus and having the drivers and horses deliver you there.[6]

The department stores on Ladies' Mile were only one type in a myriad of leisure and entertainment venues that relied on horse-powered streetcars and omnibuses to shepherd visitors their way. As noted, *The Black Crook* was playing at Niblo's Garden Theater, but there were many other theaters. Over at the Booth Theatre on Twenty-Third Street and Sixth Avenue, at the edge of Ladies' Mile, legendary actor Edwin Booth, who had returned to the stage after a hiatus forced by his brother's assassination of President Lincoln nearly a decade earlier, was starring in *Richelieu*. *A Morning Call* was playing at the nearby Wallack Theatre on Thirteenth and Broadway, *Saratoga* was being performed at the Fifth Avenue Theatre, and Lina Edwin's Theatre was staging *Hunted Down* with British actress Laura Keene, who had starred in *Our American Cousin* at Ford's Theater on the evening of Lincoln's assassination. Nearby, at Fourteenth Street between Second and Third Avenues, the *New York Circus* was home to "grand equestrian and gymnastic performances." Any of these could be reached easily by the streetcars, omnibuses, private hacks, and other horse-drawn vehicles flooding the streets of Manhattan in the early 1870s.

As residences and businesses continued their march north, uptown, more and more leisure venues could be found there. By 1858, the spectacular 843-acre Central Park with its picturesque landscapes and features, designed by Fredrick Law Olmsted and his colleagues, was opened to New Yorkers and visitors. Many wealthy New Yorkers would harness up their horses and parade down the miles of paths that were at the heart of the park. In 1859, the Central Park commission's report tallied 1,996,981 pedestrians strolling leisurely through the grounds of the park, and 709,010 vehicles (virtually all horse-powered). "Allowing the average of three persons to each vehicle," the commission report noted, "we have a grand total of 4,195,595 persons who have visited the Park during the year." By 1872, the park had long become a city treasure. Besides the wealthy New Yorkers with their own horses and buggies, and folks who lived in walking distance, the majority of the millions of pedestrians visiting Central Park each year arrived by streetcar and omnibus.

Above and beyond its attractions, Central Park was also home to the American Museum of Natural History, which opened to the public in the spring of 1871. New Yorkers had been clamoring for more museums, to put them on par with London, Paris, and other European cities, and now they had one, just an omnibus or streetcar ride away, on the second and third floors of Central Park's Arsenal Building on Sixty-Fourth Street. Plans were already afloat to build a larger, permanent home for the museum. For park visitors who preferred live animals to museum specimens, there was also the Central Park Menagerie that was home to 601 animals.

The streetcar or omnibus that brought New Yorkers to the American Museum of Natural History and the Central Park Menagerie also deposited them close to another new wonder. A ten-minute walk south of the American Museum of Natural History, the Metropolitan Museum of Art opened its doors to the public on February 20, 1872. Temporarily housed in the Dodworth Building at 681 Fifth Avenue, between Fifty-Third and Fifty-Fourth Streets, the museum was touting the coup it had pulled off in what art historians have dubbed the "Purchase of 1871." In a bold and brash move that shook the art world, a founder and trustee of the Metropolitan Museum of Art named William T. Blodgett took advantage of the jitters in the Paris art market that resulted from the Franco-Prussian war and purchased 174 paintings, many from European masters such as Anthony Van Dyke, Nicolas Poussin, and Giovanni Battista Tiepolo. The public's enthusiastic response to the Purchase of 1871 was all that museum had hoped for.

In addition to what could be found in the vicinity of the Ladies' Mile and Central Park, countless other amusements and leisure venues beckoned in New York City. Horse-powered transport could take visitors over to the Washington Square area near Fourth Street and Fifth Avenue, which had previously been used as a potter's field, and then a parade ground, but was put under control of the Department of Public Parks in 1871. Ignaz Pilate, who had worked with Olmsted in constructing Central Park, was building curved pathways surrounded by new plantings and small gathering places to create an inviting new spot offering respite from the surrounding city for New Yorkers to enjoy. Further away, for residents of Brooklyn, horse-powered vehicles were taking them to enjoy a day at the Coney Island resort: Manhattanites could join them by taking a ferry or steamer across the East River and hopping on a streetcar for the ride from the heart of Brooklyn.

When summer arrived, the streetcars were ready, running to the Coney Island resort "as regular as clockwork, every twenty minutes," the *Brooklyn Eagle* informed its readers. The streetcar companies, the paper continued, "have made ample preparations in the shape of more cars, and fine horses, with good drivers, and courteous conductors, for all of which the traveling public are much indebted." Those streetcars were usually packed to the brim, with "the beach covered by parties of promenaders who took a foretaste of the pleasure of summer in the invigorating breeze from old Neptune's realm."[7]

Baseball fans in Manhattan and Brooklyn could also board a streetcar (after a ferry ride for fans coming from Manhattan) and head over to the home of the New York Mutuals baseball team at Union Grounds in the Williamsburg section of Brooklyn. Just a stone's throw from Union Grounds, the Brooklyn Atlantics baseball team played their games at the Capitoline Skating Lake and Base Ball Ground.

So much of the everyday life of both the citizens of New York and the economy of the city relied, directly or indirectly, on urban horses. But as the leaves began to drop in the autumn of 1872, the Great Epizootic was taking on a head of steam and about to change everything.

# 5
# City Without a Horse

The citizens of New York City knew the havoc that horse flu was wreaking in Canada. On October 5, the *New York Herald* and the *New York Tribune* ran short articles on the outbreak of the disease in Toronto, each noting that seven hundred horses were sick and that the "streetcar companies and livery stables are affected." That same day the *New York Times* told readers that "an epidemic of a catarrhal nature has broken out among the horses of Montreal." Then, for two full weeks, with one exception, not a single story about the epizootic appeared in the city's major newspapers—the *Herald, Tribune, New York Times*, and *New York Sun*—and there was little evidence on the street that people were talking about it. The lone exception, which arguably proves the rule, was a one-paragraph story in the *Times*, in the paper's "Minor Items" section, about the horse disease in Toronto.

Toronto and Montreal were a long way from New York City, but this was the era of the telegraph and updates would have been arriving daily over the wires. When the press silence finally broke on October 19, the short stories that appeared in the *Times* and the *Herald* were again about the outbreaks in Canada: "The epidemic among horses is now extended throughout the Dominion," a piece in the *Times* informed readers, "seriously interrupting the business of the country. . . . The farmers in the country are unable to bring grain to market. . . . [The] disease is not often fatal but recovery is slow." It wasn't until the following day, October 20, that the Great Epizootic's arrival in the United States was reported in a New York City newspaper. The whole of that story in the *Herald* was a single sentence with a byline from Buffalo: "The Canadian horse epidemic is prevailing to an alarming extent among the horses in this city."

Newspapers in Buffalo had reported the disease a full five days earlier and telegraphic wires would have made that news available immediately. Even news of the disease reaching Albany and Newburgh no later than October 19 would have been known to New York City newspapers immediately over the wires and within a day, two at most, via the constant stream of people traveling between Manhattan and those cities, particularly the capital in Albany. Indeed, an article published in the *Herald* a few days later, on October 23, demonstrates that at least that newspaper, and so likely others, had been "trac[ing] the track of the pestilence from its first advent in Buffalo, Rochester and Niagara down to Albany and Troy, where it has been raging during the past week" and so had known about the movement of horse flu in New York State from as early as October 16. And yet no word of the disease arriving in those cities could be found in the newspapers of New York City. Rather than being in a state of high alert, the press and the citizens of New York City seemed remarkably indifferent.

Had they acted when they first learned of the epidemic, the citizens and government of New York City might have taken any number of measures, ranging from a widescale campaign to disinfect stables and rest horses to boost their resistance, up to a ban on horses entering the city from elsewhere. But they didn't, and the reasons why are hard to determine: it wasn't as if New Yorkers were talking about the disease and directly dismissing it as much as they were simply ignoring it. Perhaps those inclined toward the miasma theory of disease thought New York City might be spared the atmospheric disturbances and bad air that brought on horse flu; if they did think this, however, they left no record of it. Or perhaps the information from Canada that horse flu wasn't fatal most of the time led the residents of New York City to think of it as a nuisance rather than a serious threat. Still, they would have heard from the scant newspaper reports, as well as travelers, telegrams, and other sources that when the epizootic hit a city, it had a crippling effect on the economy, and so, in retrospect at least, their inaction seems hard to fathom. Whatever the reasons for their apparent indifference, the citizens of New York City, along with their equine employees, soon had to deal with the Great Epizootic.[1]

It was Alexandre Liautard, one of the country's leading veterinarians, who broke the bad news that horse flu had arrived in the city. Liautard was hardly the typical New Yorker. Known as Dr. Liautard to the general public and Frenchie to his friends and adversaries, Liautard had received his training at the National Veterinary School (Ecole Nationale Vétérinaire) in Toulouse, one of the first French veterinary schools, which had opened its doors in 1828, largely in response to outbreaks of rinderpest, a viral plague that devastated cattle in Europe, killing almost every animal it infected. In 1860, after a short stint in the French army, twenty-five-year-old Liautard emigrated to the United States and quickly set up a practice in New York City on Lexington Avenue. In 1863, Liautard's reputation had already reached such heights that when the first meeting of the United States Veterinary Medical Association—which, in 1898, became the American Veterinary Medical Association—held at the Astor House in Lower Manhattan, he was appointed secretary of the association and, cribbing Cicero, proposed it adopt the motto "Non Noblis Solum" (Not for Us Alone). The following year, with the blessing of the New York State Legislature, Liautard's veterinary practice was transformed into the New York College of Veterinary Surgeons, where Liautard was appointed a professor: in time, he became dean of the college.

It is from Liautard that we can determine the date that horse flu was recorded for the first time in New York City. By 1872, Liautard was the veterinary surgeon of the Board of Health of New York City. In a report on horse flu he noted that "on the evening of October 21st, only a few animals were affected." Though he didn't have the data to support it, and he generally was conservative in his approach to veterinary medicine, in a hyperbolic flourish Liautard added, "But on the morning of the 22nd I doubt if there was a single animal of the equine species which was not attacked."

In fact, on October 22, more press coverage was given to President and Mrs. Grant's visit to New York City than to the epizootic. A scant two weeks before the national election, the Grants were in town to spend time with their daughter Nellie, who was returning from a visit to Europe on the Steamship *Russia*. The sole extent of newspaper coverage about the disease on the October 22 were two *Herald* articles reporting on horse flu in Buffalo and Rochester, but not New York City itself. The longer of the *Herald*'s stories warned of a "Canadian Epidemic Spreading at an Alarming Rate," and suggested a cure using steam baths to administer ammonia "mixed with two or three ounces of honey or molasses . . . poured into the [horse's] mouth," but the tone of the article was hardly alarmist.

Twenty-four hours later, on Wednesday, October 23, the situation could not have been more different. That day saw a slew of articles in the *Times*, the *Herald,* and the *Sun*, all warning the citizens of New York City of the dire straits they were now in as a result of the Great Epizootic. In what must have struck New Yorkers like a thunderclap, the *Herald* ran a huge story titled "THE EPIZOOTIC," in which readers cast their eyes on a subheadline of "Seven Thousand Horses in the City Stricken with Catarrhal Fever During the Past Twenty-Four Hours," followed by sub-subheadlines of "Millions of Dollars in Horseflesh at Stake" and "How the Disease Will Affect Local Traffic and Travel." The reporter who penned the story began by assuring his readers that although what they were reading would seem almost impossible to believe, it was all too true. He had personally visited "all the principal stables in the city," and, if anything, the seven-thousand figure in the headline was an underestimate. No one's horse was safe: from the "squalid shed that shelters the costermonger's nag to the magnificent palace where the millionaires' thoroughbreds recline at their ease, surrounded by all the luxury that wealth can purchase," horses were falling ill at an alarming rate.

"My dear fellow, I never saw anything like it," one wealthy New York City livery owner told the *Herald* reporter. On the evening of Tuesday, October 22, he heard one horse in his stables coughing. The morning of the 23rd, he "found every horse I own afflicted with a husky, dry cough," with those nasty coughs "enough to break [their] back." And that livery owner's story was anything but unique: "It was the same story everywhere," *Herald* readers learned. "Horses that were to all intents and purposes in perfect health on Monday night [October 21] were coughing badly yesterday morning." A thousand horses in a single stable—likely the giant stable housing the Third Avenue streetcar line's two thousand horses—came down with influenza in a span of twenty-four hours. "A fiendish diligence," the *Herald* announced, had "cast loose in the closely confined limits of this city a seed of disease" (figure 5.1).

As to the cause of that disease, *Herald* readers learned that "although there are some differences of opinion as to the exact nature of the disease it is conceded by all to be a species of influenza." The question was whether that influenza was the result of an infectious or miasmic agent. Here, the *Times* story on October 23 weighed in, telling readers that "the disease is not infectious and it is stated by the veterinary profession to be in the atmosphere."

An article in the *Sun* that same day concurred. Soon the *Herald* wrote that "the air is surcharged with this epidemic," and a week later told readers of "cosmical agencies [which] operate upon our planet through its great atmospheric machinery. . . . The northeast wind which for some time has descended over our country from

**Figure 5.1**
The Third Avenue streetcar line stable during the Great Epizootic. From *Frank Leslie's Illustrated Newspaper*, November 9, 1872.

Canadian borders . . . [is] apparently charged with climatic conditions favorable to the development of influenza . . . and similar affections."

Weather reports from October of 1872 show that it wasn't especially rainy or humid that month. No rain was recorded between October 19–22 and only a drizzle on the 23rd, and so at least that "atmospheric condition" that advocates of the miasma theory most readily associated with the outbreak of disease was absent. But miasma theory was in the air in another sense, in that it was mentioned more than fifty times in the 1872 issues of New York City newspapers appearing *before* the epizootic struck: indeed, on the very first day of 1872, when reporting on the illness of the Prince of Wales, the *Sun* was quick to point out that "the almost fatal illness of the Prince shows that the deadly miasma may be as prevalent in marble palaces as in the haunts of the poor." In that case, the miasma was the result of "sewer gases," but the general point would not have been lost on readers, and if it were, they would have had plenty of other chances to read about "a constant flood of foul humidity" in the city itself, causing no end of health problems.

A burning question was how to treat ailing horses, and here there was no consensus among veterinarians. "A dozen different methods of treatment are proposed," a frustrated *Times* reporter told readers, "each of which is asserted to be merely infallible, but, in the meantime, the epidemic pursues its course, heedless of the efforts made to stem it." Some, including a few of the streetcar stable owners, liked to administer to their horses Taylors Compound, produced by the Manhattan Feed Mill Company, but elsewhere owners were force-feeding horses a mixture of tar, alcohol, and belladonna. The procurers of another advertised cure called Crumb's Carbolic Ointment even managed to get an advertisement into the same issue when the *Herald* ran the story on October 23 that first announced what a threat horse flu had become to the city.

Whatever "cure" was used, if it made horses feel any better at all, it did so marginally, and in some cases, it might have made the situation worse. A canvassing that one *Herald* reporter did of the Second Avenue, Third Avenue, Fourth Avenue, Sixth Avenue, Broadway/Seventh Avenue, Eight Avenue, Ninth Avenue, and Tenth Avenue streetcar lines' stables found many a sick horse in most of those stables, though victims were largely showing just the very early symptoms of coughing and runny noses. Even with extremely limited information to work from, there was now a palpable fear that ultimately mortality rates might be especially high in New York City (figure 5.2).

It seemed the epidemic's impacts were only going to get worse. "The results of this disease will probably be felt by all the community," the *Herald* warned. "Judging from the present aspect of affairs it is very doubtful how long our street cars and stages will be running or how soon we shall be reduced to an entire dependence upon shoe leather." In fact, during the early stages of the epidemic, some streetcar lines were already running fewer routes. "A walk from Harlem to the battery to business in the morning and return the same night may be conducive to health," the *Herald* story continued,

**Figure 5.2**
A distraught family with a victim of the Great Epizootic. From *Harper's Magazine*, November 23, 1872.

"but the larger proportion of the community will find it pretty hard work." Indeed, a week later, the *Sun* told of "thoroughfares . . . crowded with weary workers—men, women, and children—who, after vainly endeavoring to find transportation on a crowded car, turned their faces towards home and started for a walk which in many instances measured miles."

Washington politicians were also starting to get very worried that the "disease will probably be felt by all the community," but now it was the community writ large—the entire country. By late October, the horse flu epidemic had broken out in dozens of American cities. In response, U.S. Secretary of the Treasury George Boutwell sent orders to the owners and operators at major ports, including in Rochester: "You are directed, so long as the existing horse disease prevails in Canada, to allow no horses to be brought into your port or district from that country unless you are properly satisfied that they are free from contagious disease."

Boutwell's order might stem the flow of new cases coming from across the Northern border, but the "properly satisfied" exception meant infected but asymptomatic horses were still allowed in, and in any case, by this time most new instances of the disease were the result of interactions between infected horses *in* American cities. Furthermore, because the disease originated in Canada, there were no restrictions banning horses from being shipped across the southern borders of the United States. When "a large number of South American ponies arrived . . . in perfect health" in New York City, the very next day half were down with the flu. Though the United States was not blocking the importation of horses coming from south of the border, the Board of Health in Cuba took measures similar to those of Secretary Boutwell and quarantined all horses arriving from New York "to prevent the horse disease from breaking out in Havana." Ultimately that quarantine was not enough and the Great Epizootic hit Havana in mid-November and lasted about a month. "It prevailed to such an extent as to seriously interfere with all business requiring the services of those animals (horses and mules)," wrote Henry C. Hall, United States vice-consul-general to Cuba. "The greatest mortality was among hack and stage horses and the horses of the city railroad."[2]

The New York City's newspapers coverage of the epizootic reached frenetic heights by October 24, with two articles in the *Tribune*, two (along with numerous ads for cures) in the *Herald*, one in the *Times*, and one in the *Sun*, with headlines reading "THE HORSE EPIDEMIC," "THE POOR BEASTS," "OUR NOBLE HORSES DYING," and more. Almost all the major streetcar lines were suffering, some more than others. Accompanying a *Herald* reporter who went from stable to stable to assess the general situation was a man the paper dubbed "The *Herald* Horse Plague Commissioner." Before that pair began bombarding readers with data on the horse flu, they added a caveat that the owners of streetcar lines had an incentive to underestimate the extent of the damage from the Great Epizootic among their own horses. They certainly thought as much of a Mr. Roberts, the stable boss at the Bleecker streetcar line, who they described as "a horsey-looking fellow with a blonde face and a quid of tobacco, [who's] up to his shoulders in business." What came from their chat with

Mr. Roberts was the reporter's assessment that "the great trouble in obtaining information at the horse railroad depots is that the officials believe that the knowledge that a great number of horses are about to die will depreciate the value of the stock of the company."

Though Mr. Roberts of the Bleecker Street streetcar line was of little assistance, the *Herald* reporter and his partner were able to estimate, and likely underestimate given their caveat, the extent to which the epizootic had struck other streetcar lines. The superintendent of the Eighth Avenue line told them, "We have 1,000 horses on our line. . . . near as I can judge about two-thirds of our horses are slightly but not seriously affected. They are seized with this hacking cough and choking." The Third Avenue line, that applied a homeopathic remedy of syrup of licorice and molasses, had two-thirds of its nearly two thousand horses coughing and showing early symptoms, half of the Fourth Avenue line's 700 horses were sick, and the Ninth Avenue line had about a quarter of its 200 horses showing symptoms. The Broadway line got off relatively easy with less than 10 percent of their horses displaying signs of the flu. Stories in the *Times* and the *Tribune* filled in the data for other streetcar stables: the Sixth Avenue line had nearly 20 percent of its horses in their veterinary hospital and over in Brooklyn, the majority of the 2,000 horses on the City Railway line were down.

The major express companies, who moved merchandise and large items within the city and to nearby locations, were all scrambling to keep their operations going. The American Express Company had 115 sick horses, The American Merchant's Express reported that nearly all of its 150 horses were down, The National Express had all of its fifty horses showing flu symptoms, the United Express Company said half of its fleet of 100 were ill, and fifty of the 200 horses powering the Adams Express Company were out of commission.

The horses working directly for the city were just as susceptible to the epizootic as those employed in the private sector: "Many of the fire department horses are sick," the *Herald* told its readers. Indeed, the paper reported in its story titled "The POOR BEASTS," "It is calculated that from thirty to forty percent of our beasts of burden in New York are afflicted and this would on a moderate calculation give us from 8[,000] to 10,000 horses in the various stages of the malady."

In response to the outbreak, the New York City Board of Health held an emergency meeting. The report that came from that meeting, penned by Alexandre Liautard, was fairly cursory, noting the disease was almost surely influenza and that it was not especially dangerous, and was easily diagnosed. Suggested treatments included fumigation of stalls and applications of mustard to the throat: bleeding the patient, the report noted, or use of overly aggressive "arterial sedatives" and purgatives were strongly discouraged. As to the cause of the epizootic, the report took a miasmic slant, by noting that it did not appear to be "contagious and infective." As a result, Liautard "regret[ted] the order of Secretary Boutwell preventing the importation of horses from Canada." If the disease wasn't spread from horse to horse, Liautard reasoned, what good could come from such a ban. "If it is contagious, it is too late," he wrote, "if it is not, it [the order] will only increase our difficulties in the horse trade."

The *Sun's* story on October 24 captured the mood in New York City: "The chief topic in this city was the new horse complaint which had suddenly made its appearance, and has quickly sprung into alarming proportions, threatening here what it has accomplished in Buffalo and elsewhere . . . a partial or total cessation of travel and a vast injury to trade." There was general agreement that the situation in the streetcar stables would only get worse before it got better and that the economic consequences of the epizootic were only just beginning.

Products and materials were slow to make their way to the ports of New York City and then slow to move from there to their ultimate destinations intracity. The construction industry lacked both horse-power and the supplies such horse-power provided to keep the city growing at a pace commensurate with the start of the Gilded Age. Stores and shops, factories and manufacturing houses all faced similar constraints. Ice companies had no horses to deliver ice and coal companies were similarly affected. And it was still early, as the epizootic had been recorded, at most, three days earlier (though some horses might have been infected, and asymptomatic, before then).[3]

Over the next ten days, between October 25 and November 3, New York City's *Times, Herald, Tribune,* and *Sun* published no fewer than fifty articles (not including advertisements for cures) about the Great Epizootic. Reports of between sixteen thousand and eighteen thousand horses suffering flu symptoms led to sensational headlines such as "THE NOBLE HORSE IN PERIL," "THE EPIZOOTY EXTENDING: THE AWFUL FUTURE," "STRETCHED IN THE STRAW," and "PROGRESS OF THE PESTILENCE," with newspapers again reminding readers their reports on the number of infected animals might be an underestimate as streetcar companies in particular "do not seem overanxious that the truth of this particular should be announced."

Sensational headlines aside, in a city where so many relied on horses directly and indirectly for their daily keep, sick animals were forced onto the streets to pull streetcars and omnibuses, deliver merchandise, pull fire carts, and more. Some questioned the wisdom of such actions, both from a moral and economic perspective. "The bad and cruel plan of working horses unmistakably afflicted is bearing its legitimate fruit in the utter prostration of the wretched beasts," the *Herald* opined, "and will in all likelihood entail losses from death which will far more than counterbalance the petty amount gained by two or three days work out of a tottering animal." That bad and cruel plan made some see the ailing horse, forced to work, as a model of spirit in the face of adversity: "There is something heroic," the *Herald* proclaimed, "in their endurance and fortitude, by the compulsion of nature, mute and uncomplaining, by the strength of heart, which can be found even in this beast steady and unflinching."

Just days into the outbreak, New Yorkers found themselves scrambling for transport that could move them beyond the distance they could cover by foot. And even when they found a streetcar or an omnibus, it provided anything but a pleasant ride. "No amount of conductorial ingenuity could have squeezed one more unfortunate, though no bigger than Tom Thumb, into any one of the cars which took the businessmen to their homes at the Harlem end of the island," one article described it.

Though mortality rates remained low, nearly one half of the fifteen thousand horses employed by streetcar companies and various stagecoach lines were sick. "Some well-informed men in the railroad business," the *Sun* reported, "say that the present indications point to a total suspension of stage and car travel." The *Herald* chimed in that "soon New Yorkers, it is to be feared, will have to cry, with almost as much earnestness as Richard III's 'my Kingdom for a horse!'" Some were already not just crying that but also doing Richard one better and paying the downtrodden to take the place of the ailing horses: "Crowds of poor and ragged men, who have been out of regular employment for a long time were mustered and could be seen playing the part of the noble horse before wagons and hand carts." Down at the piers, "the yo-ho-ho of the longshoremen," pulling wagons "drowned the curse vented over muddy boots and bedraggled garments." When *people*-powered vehicles failed, horse owners employed oxen to replace their steeds, with the *Herald* opining, "We shall be glad to note the hour when this invasion by Taurus is on its last legs; we believe that New York does not want any more *bos* [cattle] on our streets . . . that beef can be put to better account" (figure 5.3).

The situation was deteriorating just as rapidly in the stables housing horses employed in city services. New York City Police

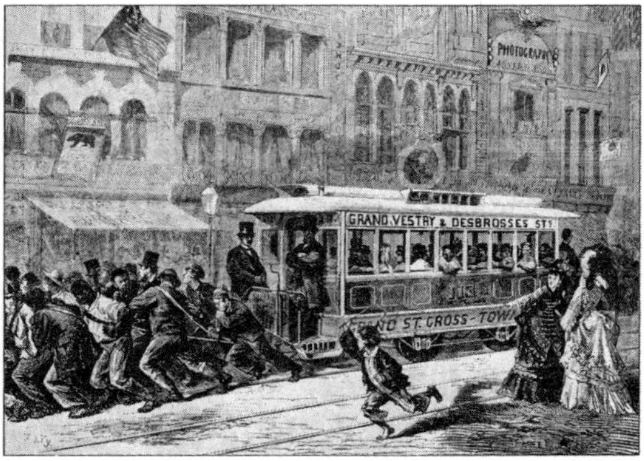


**Figure 5.3**
A team of men pull a streetcar in Lower Manhattan during the Great Epizootic. From the Brooklyn Historic Railway Association.

Department horses were pulling up sick and some newspapers reported that half the New York City Fire Department horses being afflicted. Others report that of the 154 horses the Fire Department relied on, only two were healthy: of the remainder, all suffered some symptoms and thirty-four were sick enough to be unfit for work. To keep fire carts in operation "hand companies" were organized and humans were substituted as the source of power. Without horse-drawn hearses, "long lines of walking funerals" were becoming all too common a sight. City services were also suffering. The post office "fear[ed] some derangement in the mails." New York City officials were telling residents to minimize the amount of ash they put in their garbage cans because there were scant horses to pull the collection carts, and as so often is the case, the downtrodden were hit the hardest. "In many of the poorer parts of the city," the *Sun* told its audience, "the ashes and rubbish are rapidly accumulating on the sidewalks and in the gutters."[4]

As the Great Epizootic worsened, New Yorkers began to realize just how much they relied on horses as the engines of the city. "It is only when we are threatened with the loss of his services that we appreciate the extent of our reliance upon the horse," the *Tribune* opined. "Without him we should gradually go back to the condition of Troglodytes. The transient suspension of his uses, next to scarcity of bread or an earthquake, is one of the most serious disasters which can befall a community." The *Nation* went further still: "We really ought . . . to be thankful that the present epidemic has brought us face to face with the startling fact, that the sudden loss of horse labor would totally disorganize our industry and our commerce, and would plunge social life into disorder, would threaten the lives of hundreds of thousands of human beings . . . and might expose our great cities to destruction by fire."

Many calls were made to use steam dummies—steam-powered streetcars. Alexander Bonner Latta, the inventor of the steam dummy, had experimented with his device on the street rails of Cincinnati as early as 1860, and while it worked fine, it terrified the horses pulling other streetcars. Now, as most horses were sidelined with the flu, that side effect of steam dummy engines was not nearly the constraint it had been. Steam dummies, some of which had a faux horse head at their front to mimic horse-powered cars, and others designed to look like a steam horsecar—were in demand in New York City, but only a handful existed in the city because the long-standing ban on steam railroads running below Forty-Second Street had been seen as the death knell of steam-powered cars within Manhattan.

As early as five days after the horse flu entered the city, serious economic consequences were being felt and a dread of what might follow was palpable. "The threatened consequences to the city of the temporary loss of its mode of power," the *Tribune* warned, "are really becoming not merely vexatious but serious and may yet cause great suffering." Without adequate streetcar and omnibus transportation, New Yorkers were restricted to buying goods at shops within walking distance, which was crushing sales at department stores and markets like those at Catherine Street and others all over the city. And even when people managed to get to those establishments, supplies were low, as the supply chain into the city had all but shut down. Steam railways, of course, could make it to the depots above Forty-Second Street, but these depots were daunting spectacles of huge piles of merchandise of one sort or another waiting to be delivered to shops and warehouses throughout the city. "The freight depots of the various railways are crowded with bales, boxes, casks and packages of every description," read a *Times* story on October 31, "the consignees having no means of removing them." It was even worse at wharves. The epizootic elsewhere in the state was shutting down shipments along the Erie Canal. Those boats and ships from upstate and other states and countries that did find their way to the docks in New York City immediately discovered that there were no horses to either unload ships or move goods into the city. "Steamship companies are suffering greatly from wanted animals," read one account. "Immense piles of goods lie waiting transportation on the piers. Steamers arriving here find considerable difficulty on unloading."

Compounding the economic turmoil was the problem that docked ships could not leave until their cargo had been unloaded and new cargo was brought onboard for their return voyage, which meant that open slots for other vessels that managed to make it near the docks were virtually nonexistent. As a result "large sums have been paid for demurrage for several days past," the *Herald* told readers. The price to move cotton from the wharves to the warehouses, for example, had quadrupled from twenty-five cents to a dollar a bale. "Several grain dealers had each two thousand barrels of flour still lying on the wharves which had lain there for several days," one grain merchant told a reporter from the *Sun*. Because of the difficulty getting goods into the heart of the city, that merchant was already paying $40 a barrel for flour, when the normal price was $10, and now much of that very expensive flour was stuck onboard a ship for lack of horse power to get the barrels into a warehouse and then to market.[5]

Cotton, coal, and so much more remained onboard ships, large and small, with perishable food items, particularly fruit, beginning to rot. Farmers just outside New York City lacked the horse power to bring their crops into the city. The prices for these suddenly rare staples soared on the open market. Trades of all sorts were also stymied. "The dry goods trade is heavily imploded," the *Sun* informed readers, adding that "many who take ice and milk find [also] their supply tardy." The very sounds of the city were changing: "The rattle of the milk cart and the wild welp of the milkman," as one newspaper article put it, "no longer disturbed the morning slumbers" of the residents of New York City. No sector of the economy seemed immune to the effects of the Great Epizootic. Even real estate values were hit, as people became leery of purchasing land in surrounding areas like Long Island and Westchester County when transportation there was no longer a given.

The epizootic struck the amusement and leisure industry as well, largely through access to streetcars and omnibuses. In some cases, such as the baseball championship games between the Baltimores and the Mutuals being played in Williamsburg, Brooklyn, where much of the crowd would normally have commuted from Manhattan to partake, "the attendance was small," the *Herald* reported on October 30, "owing probably to the fact that the means of reaching the place of playing were very limited, most of the horsecars having been withdrawn because of the epizootic." Horse races too were cancelled, both because some trotters were ill, but also because of the difficulty of getting to racetracks.

The effect of the horse disease on attendance at musical performances and plays varied from venue to venue. Many venues saw audience size decrease, sometimes markedly, but at Steinway Hall, where Anton Rubinstein and Henryk Wieniawski were performing, attendance was respectable, "considerin' the horse disease," while a rehearsal performance of the New York Philharmonic at the Academy of Music drew a full house from those in walking distance. The epizootic had effects above and beyond those on attendance. Horses were actors of a sort as well and "the two very valuable snow-white horses engaged in the Pompeiian scene at the Grand Opera House," the *Herald* told its readers, "have been stricken down by the prevailing distemper."

On October 31, a rough estimate of the total loss to the New York City economy from the start of the outbreak of the Great Epizootic lay at $6 million, which amounts to approximately a modern-day equivalent of $155 million. The only people who seemed to profit from the disease were pickpockets, who took advantage of the fact that the few streetcars and omnibuses that were running were jam-packed, making their job all the easier, and price gougers, who understood all too well that when a commodity is scarce in a large market, extortionist prices are sometimes paid. Hack drivers with uninfected—or sometimes infected—horses were charging a whopping $10 to take Wall Street bankers and brokers up to Twentieth Street: "The hackmen, with characteristic rapacity," the *Herald* reported, "have tripled the fares which they demand of passengers." And it wasn't only on Wall Street where people were being hustled. When a *Sun* reporter asked a merchant, whose company relied on horses to transport their goods, whether those with healthy horses were price gouging, and if so, by how much, the man replied "from 50 to 100%. I have heard that they ask more than that on both the east and west sides." Other merchants told the same reporter that price gougers were tripling the price of delivering honey from piers into the heart of the city.[6]

By November 4, 1872, the day before national elections, and less than two weeks after the outbreak, New Yorkers were reeling from the impact of the Great Epizootic on every facet of life in their city. And there was more to come.

# 6
# Sorry Beasts That Melt Away into Melancholy Grease Spots

The results of the presidential election of 1872 were a foregone conclusion: Republican incumbent and Civil War hero (at least in the North) Ulysses Grant easily defeated Democratic challenger Horace Greeley, editor of the *New York Tribune*. With regard to the election and the Great Epizootic, when the first major stories about the horse flu hit the pages of New York City newspapers on October 23, the number of Democrats, Republicans, and Liberal Republicans in New York City who had ever heard the word "epizootic," let alone knew what it meant, was surely no more than the number of veterinarians and physicians in the city and a handful of others. Fewer than ten days later, not only was the word "epizootic" bandied about on the streets, but it had also migrated into political parlance.

"MANIFESTATIONS OF LIBERAL EPIZOOTIC" read one headline of the November 2 issue of the *New York Herald*. The agenda of the liberals, like Greeley, the paper opined, raised "a question worth considering by the medical faculty," namely "whether the violent storm or fierce denunciation and rhetorical filth now sweeping the state is not the real cause of the shaky condition of that intelligent animal, the horse."

This was not the first time the language of disease had been commandeered for political purposes. History is full of other examples of medical terms eventually making their way into conversations of laymen, but what was new was the speed at which the word "epizootic" went from being uttered by a few medical professionals to making the political headlines. And there was more to the epizootic connection to the election of 1872 than just breaking into political vernacular. Some New Yorkers thought the impact of the epizootic on voting patterns favored Greeley. "The rich may ride in chaises and the poor may walk by—but now the men of millions and the men of pennies have to foot it together," the *Herald* wrote on October 26. "Highty tighty Republicans will not walk through the mud to the polls but Democrats will wade out in their brogans to vote. . . . [The] 'epizootic' [is] a providential affair all together and designed by a 'higher law' for the comfort and encouragement of good liberals. . . . It will give ten thousand additional votes to Greeley." It's impossible to know if that happened, but we do know that concurrent with an increase in population size per se, in every presidential election between 1848 and 1908, *except for the election of 1872*, the number of votes cast in New York State increased. There was no lack of interest in the presidential election that year, particularly in New York, what with Greeley being the editor of the *Tribune*, and so it seems reasonable to assume that the dearth of transport to polling stations as a result of the epizootic was one, or perhaps, *the* primary factor explaining voter drop-off in 1872.[1]

As the Great Epizootic was devastating New York City, both action and causation were being debated. What was causing the disease? Who had the expertise to help? What actions should be taken? On actions, there were two related questions: first, what specific treatments should be applied to ailing horses by those deemed to be in a position to help? Second, and the larger philosophical question, was the goal of treatment to fix a machine (albeit a biotic one) that powered the engine of city life, or were horses, indeed all animals, due some level of protection and sympathy because they were living, breathing creatures?

On cause, the miasma theory for the origin and spread of the disease remained the overwhelming consensus among veterinarians and stable owners as the cause of their horses' woes. It was "the fetid quality of the atmosphere—undoubtedly a gaseous problem" that was to blame, stated one *Herald* story. As a result, for the overworked New York City streetcar horses, there was little concern that they were especially susceptible to disease and that the horse next to them as they pulled their load might be infected and further spread the disease.

When horses began to show symptoms, the veterinarians and others who cared for the horses in stables did not isolate or quarantine them because they did not believe the disease was spread from one animal to another. "The 'hospital' branches of every car stable in the city are filled," read a *Times* article on October 24, "and in many cases the affected animals are kept in their usual stalls from lack of accommodations elsewhere, which, it may be as well to mention, is not regarded as dangerous to the unafflicted animals as the epidemic is not believed to be infectious."

The *Herald* published numerous theories on what the miasmic agent causing the disease might be. One article titled "The Medical Geography of the Equine Epidemic" opened by informing readers that "the great horse epidemic . . . seems to be intimately connected with certain well marked meteorological phenomena." Had people just been paying attention to the scientific writing on the wall, the *Herald* writer continues, and counted the twelve-year cycle of sunspots, they would have known this blight was coming. "Solar researches . . . have demonstrated the close thermal and atmospheric relations of our globe with the recurrence of sunspots . . . in period[s] of nearly 12 years . . . which brings the greatest area of an electrical derangement to our planet." 1848 had seen a burst of sunspots, as had 1860, and now in 1872, read the *Herald* story, one should naturally expect "numerous vegetable blights and the present scourges of the animal kingdom." Another article in the *Herald* proposed a rather different miasmic agent. Now, the epidemic, it was proposed, "may be caused by noxious exhalations from the soil [and] due to a warm and damp atmosphere. The inhalation of this miasma might act on the equine tribe after the fashion of this disease, just as certain soils are sure to produce fever and ague in human beings."

Cornell professor James Law, one of the most respected veterinarians of the day, took a vocal position against the miasma theory. Law was born in Edinburgh, Scotland, on February 13, 1838. At the tender age of sixteen, he enrolled in the Edinburgh Veterinary College, where Professor John Gamgee, an expert in diseases in horses and cattle, became Law's friend and mentor. Three years later, in 1857, Law graduated with honors and then spent a year at the University of Edinburgh's Medical School, where he took courses with Joseph Lister, a leader in the movement to use antiseptics during surgery. Following that, he spent time studying at veterinary colleges in Lyons and Alfort, France. In 1861 Law returned to Scotland and began teaching at the New Veterinary College in Edinburgh. At that time, Scotland was a hotbed for veterinary science. The United States was not. Ezra Cornell, a New York state senator, and others wanted to change that.

In 1865, Cornell founded his eponymous university, after he and Andrew D. White, a fellow state senator, pushed through a bill for a state-funded land grant university. Cornell instructed White, who became the university's president, to head to Europe to recruit professors, including one to fill a chair in veterinary medicine. It took time before that recruitment trip got underway (Cornell University did not open its doors until October 7, 1868), and as it was being mapped out, Law's mentor, John Gamgee, was in the United States at the request of the U.S. Department of Agriculture, to study an outbreak of pneumonia in cattle. Gamgee paid a visit to Cornell and White and when he learned of their hunt for a veterinary chair, he told them James Law was the man for the job.

In the summer of 1868, White, at last on his recruitment visit overseas, was learning just how abysmal American veterinary medicine really was. "There is at present hardly anything in the way of scientific veterinary teaching in America," White wrote Cornell in late June, after spending time at the universities in Europe, "indeed I may safely say there is none and to build up that department is acknowledged generally among us to be of the first importance." White was overstating his case, but only slightly. The state of veterinary education in America was indeed abysmal, with many people calling themselves horse doctors and veterinarians having no formal training at all.

Over the course of the 1850s and 1860s, a number of attempts had been made to remedy this through private veterinary colleges in large metropolises. The Veterinary College of Philadelphia had been established in 1852, but closed in 1866, and it is unclear how many, if any, students graduated: the college was sold to Robert McClure, who rebranded it the Pennsylvania College of Veterinary Surgeons, and operated what one historian of medicine called a "diploma mill."

Bostonians too tried their hand at formalizing the study of veterinary medicine. The Boston Veterinary Institute opened in 1854 or 1855 but closed just five years later. A similar fate befell the Veterinary Institute of Chicago, which began admitting students in 1864, but shut down in 1869 (or thereabouts). The one veterinary program that did operate when White went on his professorial hunting trip—Alexandre Liautard's New York College of Veterinary Surgeons—was in New York City itself (rather than upstate Ithaca, where Cornell was located).

While he was in Europe, White had not forgotten Gamgee's recommendation, and a week after his letter to Cornell on the state of

American veterinary medicine, his travels took him to Belfast, where Law had opened a private practice. "I have looked through the principal Agricultural and Veterinary Colleges of Europe before arriving here," White wrote Cornell the day after that meeting, "[and] I have found several excellent candidates, but I find Mr. Law vastly their superior. He has published books and articles which have given him a high reputation on this side of the water and personally he is everything we could desire. . . . He cannot fail to succeed." The Board of Trustees at Cornell agreed, and a week later they offered the position to Law. "[The] Hon. Ezra Cornell just received from [the] Hon. Andrew D White, president of the Cornell University," the *Times* announced on July 17, 1868, "the engagement of . . . James Law as Professor of Veterinary Medicine and Surgery. . . . The professional reputation of Professor Law is of the highest character."

On August 7, James Law and his family set sail for America. Soon, both from behind a lectern at Cornell University and out in the experimental barns around the university, Law was in full motion at his new home institution. In 1871, he designed a four-year bachelor of veterinary science (B.V.S.) degree. Law soon amended the B.V.S. degree, such that two additional years in the program would yield a student a doctor of veterinary medicine (D.V.M.). In 1876, the first such degree in the United States was awarded to Cornell student Daniel Elmer Salmon—for whom the bacterial genus *Salmonella* is named.

On November 3, 1872, in the midst of the Great Epizootic, the *Herald* alerted its readers that James Law, professor of veterinary medicine at Cornell, would be presenting a lecture on the "horse disease" at 8:00 p.m. the following evening at the International Academy of Science located at 18 Cooper Union, near Sixth Street and Third Avenue in Manhattan. That next evening, after the chairman of the Academy noted that Law's talk "would be one of a series of lectures instituted by the Academy for the purpose of eliciting a livelier public interest in science," Law took the podium. No written record of his lecture was published, but one reporter noted that Law's presentation was "technical and not of the highest popular public interest," and began with a historical overview of influenza in horses, and more generally, in animals writ large.

After the history lesson, Law appears to have discussed the symptoms associated with influenza in horses and moved on to the question of the cause of the Great Epizootic. A reporter for the *Cleveland Leader and Morning Herald* who attended the talk noted that Law argued that the disease "could not be the result of conditions in the atmosphere for these could not continue untouched for the length of time that the disease runs: nor to gases or ingredients in the air, for these must be diluted and pass away; nor could the temperature be the cause, for it had occurred under a great variety of thermal conditions." As to what the cause of the epidemic was, Law did not yet make a case for germ theory per se, but rather, for the moment, remained silent on that matter. But the fact that Law spoke out against miasma theory, while Andrew Smith in Canada remained one of its most vocal proponents, and Alexandre Liautard too lined up on the side of miasma, is important in itself. Law, Liautard, and Smith were three of the most renowned veterinarians in North America. That these men disagreed on whether the Great Epizootic was or was not caused by miasma highlights that miasma theory was far from a fringe position taken by horse doctors and then amplified by newspaper reporters, but rather a serious scientific theory for the cause of disease being debated the finest veterinarians.[2]

In addition to the effects of the miasma theory, and the recommendations that followed from it, the mistreatment of infected horses was also feeding the flames of the Great Epizootic. "It was not uncommon along the streets of the city," the *Times* told readers, "to see horses dragging along with their drooping heads and at intervals coughing violently." The *Herald* wrote much the same, reporting that "the hack horses on the stands cough[ed] with painful exertion as they stood patiently awaiting the orders of the driver who would never let them rest as long as there was a chance of earning a dollar." It wasn't just the horses powering hacks, but the horses powering everything: "Of the horses that were working yesterday on the public thoroughfares," the *Herald* estimated that "fully one half should have been in the stable undergoing treatment."

From the moment that the horse flu struck New York City, Henry Bergh and members of the American Society for the Prevention of Cruelty to Animals (ASPCA), which Bergh founded, did what they could to help those sick horses. The ASPCA's work beginning in 1866 got the matter of animal protection—which in time would evolve into animal rights—onto the radar of some Americans. But it was the members' tireless efforts defending horses during the Great Epizootic that drew national attention to the ASPCA.

A native New Yorker, whose father had been a shipbuilder often employed by the U.S. government, Henry Bergh was described as "a tall, elegant figure [with] a high forehead, on which a single lock curls gracefully down in the Ciceronian style . . . his otherwise classical features are modernized by a moustache." In 1862, he was appointed Secretary of the Legation in St. Petersburg, Russia. Before he assumed that position, one historian of the animal rights movement described Bergh as "a cultured swell who had previously expressed only modest public-mindedness."

Bergh's outrage at animal cruelty was sparked in 1863, after he witnessed a Russian carriage driver beating his fallen horse. Bergh had seen drivers in Spain whip their horses to force them to move faster, but to beat an injured horse was more than he could stand by and watch. Bergh stepped in to aid the unprotected animal, and from that time on, the *Times* reported many years later, "He began there that active interference on behalf of the rights of animals to kind treatment which has earned for him a worldwide reputation." The following year, Bergh resigned his post in Russia for health reasons and on his trip back to America stopped over for an extended visit to Europe. While he was in England he met and befriended Dudley Ryder, 2nd Earl of Harrowby, who was serving as president of the Royal Society for the Prevention of Cruelty to Animals. Ryder was more than happy to work with Bergh on plans to create an American version of his beloved RSPCA.

The coming-out party for the nascent ASPCA was the "Statistics on Cruelty to Animals" lecture that Bergh presented on February 8, 1866, to the *American Geographical and Statistical Society* at Clinton Hall in Lower Manhattan. The evening of the talk was one of those dreary February nights New Yorkers knew well, but the turnout was good. Bergh began by telling the audience that "he had been impelled to the course he had adopted by a deep sense of the importance to society of the practice of humanity to 'these mute servants of mankind,' as well as justice to the creatures committed to our care and mercy by the Most High." From there he told tales of the mistreatment of animals he had seen in Russia and across Europe and drew, as the *Sun* reporter attending the meeting noted, "a dismal picture of the moral effects which bull fighting, dog fighting and cock fighting, beside numerous other modes of cruelty practiced, have had upon the communities." Bergh also told the audience of his interactions with the Earl of Harrowby and what the Royal Society for the Prevention of Cruelty to Animals had accomplished in England. A *Herald* reporter who also attended told readers that "all pronounced the lecture a success."

While Bergh prepared for the American Geographical and Statistical Society meeting, he had drafted a short petition, considered the founding document of the ASPCA:

> **Patrons of a Society Proposed to Be Founded for the Prevention of Cruelty to Dumb Animals**
>
> The undersigned, sensible of the cruelties inflicted upon dumb animals by thoughtless and inhumane persons; and desirous of suppressing the same—alike from considerations affecting the moral

> well-being of society, as well as mercy to the Brute creation: consent to become patrons of a Society having in view the realization of these objects.

By the time Bergh stood up to speak at the American Geographical and Statistical Society, this document had already been signed by the New York City mayor John Hoffman, along with Horace Greeley; John Jacob Aster, former Secretary of the Navy; George Bancroft; philanthropists Peter Cooper and James Lenox; businessmen such as A. T. Stewart, and two dozen others. While there is no record of who in the audience added their signature that night, no doubt some did, and soon the document had been signed by "one hundred of the best men in New York." Years later, while Bergh was visiting Philadelphia, recounting the early days of the ASPCA, he called this document "a species of Declaration of Independence for those who believe as I do, and I firmly believe that the day will come when it will be held in almost as much veneration as the time-worn and honorable document so carefully preserved in your State House."

Soon, Bergh, the first president of the ASPCA, expanded on his personal views on cruelty in general: "I know of nothing in nature, in reason, or in revelation," he said at an early meeting of the ASPCA, "which obliges us to suppose that the UNALIENATED rights of a beast are not as sacred and inviolable as those of a man; or that the person who wantonly commits an outrage on the life, happiness, or security of a BIRD, is not as really amenable, at the tribunal of Eternal Justice, as he who wantonly destroys the rights and privileges, or injuriously takes away the life of one of his fellow-creatures of the human race." The first annual report of the ASPCA makes it clear that society as a whole strove for "an observance of tenderness and humanity towards the inferior dependents of mankind."

The petition, Bergh's comment at that early ASPCA meeting, and other of his writings and lectures, make it clear that while one mission of the ASPCA was indeed to prevent animals from being subjected to human cruelty, there was also another moral factor at work: *people* who are cruel to animals are a clear and present danger to *human* society. And societies that sit by idly while such people go unpunished decline. In Bergh's speech to the ASPCA in which he invoked "the tribunal of Eternal Justice" for acts of cruelty against animals, he also tells his fellow members to ask themselves this: "Why should a man . . . who without the least remorse torments and destroys other creatures, hesitate to do the same, after such an apprenticeship, to any human being that happens to stand in the way of his caprices or ambition?" Bergh was convinced such a man would not hesitate and was "persuaded that the obduracy of heart and unrelenting disposition which we discover in some men, is owing in a great measure to the early indulgence of such a disposition towards the inferior animals."

Many of the ideas espoused by the ASPCA were not new in America and certainly many people before Bergh were concerned about both the cruelty to animals and the humans who inflicted that cruelty. In the 1641 Massachusetts Body of Liberties, Liberty Number 92 dictates that "no man shall exercise any Tyranny or Cruelty towards any Bruit Creatures, which are usually kept for the use of man," and while there is some question as to the extent to which this was enforced, the very liberty itself is worth notice.

In 1788, in the case of *Respublica v. Teischer* the Pennsylvania Supreme Court upheld a lower court conviction of a man for "maliciously, wilfully [sic], and wickedly killing a horse." Supreme Court Justice McKean ruled that "whatever amounts to a public wrong may be made the subject of an indictment." The Court's ruling in *Respublica v. Teischer* showed the breadth of public wrongs included not just the killing of horses, but also "the poisoning of chickens; cheating with false dice; fraudulently tearing a promissory note . . . breaking windows by throwing stones at them . . . and the embezzlement of public monies."

In large part because of the power wielded by the people who signed Bergh's petition, on April 10, 1866, the New York State legislature granted the ASPCA its charter. Section 7 of the charter reads: "The police force of the city of New York, as well as that of other places where police organizations exist, shall, as occasion may require, aid the society, its members, or agents, in the enforcement of all laws which are now, or may hereafter be, enacted for the protection of dumb animals." The ASPCA charter also included a means by which enforcing laws could fund the society, as Section 8 called for "one half of the fines collected through the instrumentality of the society, or its agents, for violation of such laws shall accrue to the benefit of the said society." Indeed, in part to generate the funds to keep the society afloat, the ASPCA in its first annual report included a section entitled "Suggestions for the Guidance of Persons Desirous of Prosecuting Individuals Guilty of Cruelty to Animals."

One anticruelty law had been on the books in New York for going on four decades when the ASPCA was formed. That law ordered that "every person who shall maliciously kill, maim, or wound any horse, ox, or other cattle, or sheep, belonging to another, or shall maliciously and cruelly beat or torture any such animal, whether belonging to himself or to another, shall upon conviction, be adjudged guilty of a misdemeanor." But Bergh lobbied for even stricter laws, which he and the society soon got passed. On April 12, 1867, only a year after the ASPCA was chartered, the State of New York passed *An Act for the More Effectual Prevention of Cruelty to Animals*. Broadening what forms of cruelty to animals were punishable, Section 1 of that act declared, among other things, that "if any person shall over-drive, over-load, torture, torment, deprive of necessary sustenance, or unnecessarily or cruelly beat, or needlessly mutilate or kill . . . any living creature, every such offender shall, for every such offense, be guilty of a misdemeanor." Bergh's fingerprints are all over this law.[3]

Many in the public and the press supported Bergh and his society. When Bergh lobbied to have the ASPCA chartered in 1866, the *Times* stepped up to his defense in an article titled "THE RIGHTS

OF ANIMALS." "We do not suppose there could be any opposition to the passage through the legislature of the bill now pending for the incorporation of the 'Society for the Prevention of Cruelty to Animals,'" the paper decreed. How could there be, the *Times* asked its readers, given the "extent and diabolical character of the cruelty that consists in running omnibuses and [street] car horses until they drop and then leaving them to die in the streets."

Bergh and the ASPCA also had vocal dissenters aplenty, including some policemen, who thought their time was better spent on more pressing matters than how people treat animals. While most policemen with such inclinations would have been discrete about ignoring just the behaviors that the ASPCA was interested in stopping, one bolder officer, Samuel Morrison, was found guilty of "inciting a mob of seventy-five to one hundred persons, by speech and manner, to resist the enforcement of the law in the case of three trucks overloaded with hay, and refusing to make an arrest when requested by Mr. Bergh."

The press too had its skeptics. Not all of the early coverage of Bergh was nearly as flattering as the *New York Times* "THE RIGHTS OF ANIMALS" piece of 1886. Bergh was often portrayed as a busybody, holier-than-thou do-gooder, and many complained "at the inconvenience which [Bergh and the ASPCA's] benevolence was causing them." As a result, he was often satirized in the city papers. "There was a terrible storm of indignation rising against the Park Commissioners for their opposition of the Marionettes, Punches and Judies and other amusements in the Park," a biting piece in the *Herald* stated on May 14, 1868. "The mass meeting of three or four hundred thousand of young New York which we proposed to call will certainly be held, aided by the mamas and presided over by Mr. Henry Bergh, if the commissioners do not abandon their cruel opposition to the wishes of the dear little folks."

Above and beyond the sarcasm, many in the press, in the public, and even among Bergh's friends thought that his call for the use of public whippings for *people* who were guilty of extreme forms of *animal* cruelty was indicative of a man who cared more for animals than for people. Bergh not only argued for such punishment, but also once told a reporter that "nothing improves ruffians so much as a good healthful clubbing well laid on," leading some to counter that the founder of the ASPCA practiced a "malevolent benevolence," and was "more gratified by the punishments he inflicts than by the good he is supposed to accomplish. . . . He has the heart of a fanatic of the Spanish Inquisition."

There were also legal challenges of all sorts to the authority of the ASPCA to enforce anticruelty laws. An early such challenge came in 1868 in the case of *People v. Tinsdale*. Bergh and the ASPCA had brought a case against Arthur Taggart and George W. Tinsdale, the driver and the conductor, respectively, of a streetcar that was running a route on Fulton Street in Lower Manhattan. A police officer that had been involved when Bergh apprehended Tinsdale and Taggart testified that the "car was unusually crowded . . . the horses slipped twice." The defense countered that the driver and conductor were employees of the streetcar company and as such they had no say on which horses were selected that day and they had no intent to harm any animals. They were simply doing their jobs, the argument went: it was the company that should be held responsible. In a major victory for Bergh and the ASPCA, the court held that Tinsdale and Taggart were responsible and fined them $250 each, ruling that "the commission of the offense under the statute does not rest upon the question of intent. The intent is assumed from the act itself." For all practical purposes, this ruling meant that Bergh and his agents could move against those committing acts of cruelty without worrying what the intent of the offender was (figure 6.1).

Bergh and the ASPCA were interested in more than cruelty to horses. They set their sights on how cattle were transported, how slaughterhouses were run, the practice of vivisection, what the city did to stray dogs and cats, the treatment of animals in zoos and

**Figure 6.1**
The ASPCA's seal (designed by Bergh and drawn by Frank Leslie), 1867. From *American Society for the Prevention of Cruelty to Animals: Annual Report* (New York: ASPCA, 1873).

menageries, and even how pigeon-shooting matches were operated. Bergh opposed cockfights and went after Kit Burns, a former gang leader of the Dead Rabbits—one of the worst gangs in Five Points—after witnessing the dog fights that Burns ran in a shady saloon called the Sportsman Hall. Bergh even got into a prolonged dispute with P. T. Barnum about Barnum feeding live rats to his snakes as part of the show at his American Museum (figure 6.2).

The ASPCA's protection of horses was special because of the unique role horses played in the city. In 1867, the same year the ASPCA was founded, the *Herald* implored Bergh to come to the aid of the horses pulling the city's streetcars. Years later, the *Times* told tale of an event that likely occurred as Bergh's agents began their work on behalf of horses in 1867 or 1868: "There are many people who can recall the first instance when he [Bergh] set its [the ASPCA's] machinery in motion," the article noted. "One blustery winter's night about 6 o'clock, when the slush in the streets was ankle deep and sleet and hail filled the air, he, with a number of his men, concentrated their forces at what was then Chatham Street, [where] half a dozen horse car lines meet, and . . . gave orders to take from the tracks every horse that had an ailment or a sore. The

**Figure 6.2**
Henry Bergh stops a crowded horsecar to check tired-looking horses. From *Harper's Weekly*, September 21, 1872.

condition of the wretched animals attached to the cars at that time was notoriously bad, and the consequence was that the society's men found plenty of work, which virtually resulted in a blockade of the lines for several hours." As a result, "thousands of people had to foot it up town in the rain and the slush, growling, cursing, hungry, wet, and mad." They were understandably agitated: "'Who did this?' was the angry question asked on all sides," the *Times* wrote, "and to this came but one reply, 'Bergh.'"

Horses played such an integral role in powering New York City that even before the Great Epizootic struck, most of the cases that the ASPCA brought to the courts involved horses there. All in all, from its inception through 1872, more than half of 1,533 cases the ASPCA took to court involved drivers forcing sick or injured horses to work. Just as important, the society tried to help the horses that fell victim to these crimes. Starting in 1867, the ASPCA began running the first horse ambulances in the country. These ambulances, with slogans like "The Merciful Man Is Merciful to His Beast" painted in eye-catching colors on their sides, serving as flashing lights do on modern-day equivalents, responded to hundreds of

**Figure 6.3**
An ASPCA ambulance for injured or sick horses. *Frank Leslie's Illustrated Newspaper*, July 24, 1869.

calls a year and brought ailing horses to veterinarians to be treated. All of which is to say that when the Great Epizootic broke out, the ASPCA was ready to protect the rights of the sick horses and to prosecute those who thought that working a sick horse was just part of doing business (figure 6.3).[4]

Bergh and his agents were on the streets from day one of the Great Epizootic, checking to see whether horses were sick or not, and whether they were being forced to pull overly crowded streetcars and omnibuses. The events of October 30, 1872, capture nicely what became a daily occurrence. That day, in the heat of the epidemic, Bergh and two officers from the ASPCA stationed themselves near a switch that enabled streetcars from the Third Avenue line to move from uptown to downtown tracks. Two policemen stood by to keep order. When a Third Avenue streetcar came along, Bergh and his team stood in front of the car, forcing it to stop, and then inspected the horses. "Mr. Bergh pointed to the yellow looking noses of the horses," reported the *Herald*, which had taken to calling Bergh "the great philohippic," and told the driver, "These animals are too sick to work. Take the other track and go back to the stables.

I'll go with you." The driver balked, but "a glance at the policeman's baton persuaded him" to do as Bergh ordered. The passengers were astounded when they realized that the streetcar was heading back to the depot and begrudgingly deboarded. Bergh then met with the president of the company that owned the Third Avenue line, who implored him not to interfere with traffic that day and guaranteed that from the next day on, they would not use sick horses on the line. Bergh agreed and when asked by a reporter how his meeting went, he summarized his day's work: "I have simply done what I can see to be my duty. I was obliged to send home, [all in all] 60 teams which were unfit for work, all of them were running at the nostrils and some were too weak to be of the slightest service." The following day he was back, this time at the Bleecker Street line, to inspect its horses.

Bergh had no qualms about pushing the envelope regarding the legal authority of the ASPCA, as he did in his "Sabbath of the Steeds" letter, published in the *New York Herald*. Though neither the ASPCA charter nor the laws of the city of New York allowed the ASPCA to act preemptively, on Saturday, November 2, 1872, Bergh wrote a short letter to the president of one of the streetcar lines: "Sir, tomorrow will be the Sabbath which the Supreme Ruler has declared shall be one of the rest of man and beast. Out of respect for this ordinance and from the considerations of mercy and humanity to the toiling brute let there be a total cessation from travel by [street]car, carriage and other vehicles on that day. Twenty-four hours of repose to the sick and weary horses will turn eminently to the pecuniary and moral profit of their owners. Let it, therefore, be shown tomorrow who are the merciful and who are the avaricious among the proprietors and public." The streetcars ran that Sunday anyway, but, not surprisingly, streetcar companies began filing injunctions to stop Bergh from shutting down their operations. Those cases got caught up in the courts and were not settled until after the Great Epizootic, in 1873, when it was ruled that the ASPCA could no longer make routine inspections to assess the health of streetcar horses, but did have the authority to arrest a driver if a "plain violation of the Statute occurred in their view."

Certainly, there were many residents in New York City, as well as in Boston and Philadelphia, where local ASPCA branches had been formed (we'll return to those societies soon), who knew of Bergh and the ASPCA before the Great Epizootic. But the actions of the ASPCA in New York City between October 21 and the end of November 1872 drew national attention and brought the matter of animal protection to the country's notice. During that five-week time span, more than a thousand newspaper articles around the country wrote of Bergh or the ASPCA, or both, and the vast majority of the articles were about their role in New York City during the Great Epizootic. Dozens of articles appeared in the newspapers in California, Illinois, Indiana, Iowa, Kansas, Maine, Maryland, Missouri, Montana Territory, Nebraska, North Carolina, Ohio, Oregon, Tennessee, Wisconsin, Vermont, and Washington, DC. Many, though not quite as many, also filled the pages of newspapers in every other state and territory in the Union.[5]

Within a year from the formation of the ASPCA, the *Herald* and the *Nation* were making the case that an analogous institution devoted to preventing cruelty to children should be formed. From time to time, citizens of New York City would bring a case of child cruelty to the ASPCA to prosecute, but, initially, Bergh and the ASPCA's council Elbridge Thomas Gerry, the grandson of Elbridge Gerry, who served as one of James Madison's vice presidents, shied away from such cases. For one thing, they had their hands full with matters of animal cruelty. For another, New York City was home to the Children's Aid Society, which focused largely on abandoned children and orphans. Founded in 1853 by philanthropist Charles Loring Brace, the Children's Aid Society was best known for its controversial "orphan trains" that moved abandoned children and orphans out to the West to help pioneer families there and instill a family and work ethic in the children. Some thought the orphan train movement a win-win situation for the children and pioneers; but others, including former abolitionists, felt that these children were being sent into a kind of indentured servitude.

Bergh and Gerry made an exception to their "animal cruelty cases only" rule in the summer of 1871, when they learned of a young girl named Emily Thompson who was being mistreated while living with her guardian near the ASPCA headquarters. Not much is known about the legal case that followed, but it did, at the very least, set an ASPCA precedent, which was followed up a few years later when, in 1874, Bergh and Gerry took up the cause of another child named Mary Ellen Wilson.

A voluntary worker from a nearby mission told them that she had learned that Mary Ellen was being severely beaten and forced to remain in her home. In part because he was angry that the Children's Aid Society had not acted when given this information, Bergh sent one of the ASPCA's agents posing as a census worker to Mary Ellen's home and confirmed what the mission worker had reported. In time, the case worked its way up to the Supreme Court of the State of New York, which removed the young girl from her home and eventually placed her with the mission worker who had first brought Mary Ellen's plight to light.

Bergh and Gerry were so moved by all they learned of the Mary Ellens that fell through the cracks in New York City's very porous child welfare safety net that in late 1874 they joined with John D. Wright to form the New York Society for the Prevention of Cruelty to Children (NYSPCC), the first such agency in the country. A reporter from the *Times* who was present at the first meeting of the NYSPCC noted that "there was a very large crowd attendance, three fourths of whom were ladies and all evinced the deepest interest in the proceedings." Wright, the first president of the society, told the audience that "children have rights which their parents and guardians should respect," and Gerry, council for the NYSPCC (as well as for the ASPCA), "alluded to the case of little Mary Ellen who was beaten with a raw cow hide for three months . . . and he predicted that as soon as the habitual abusers of children learn there is a law to reach them there will be very few cases like that of Mary Ellen." For his part, Bergh spoke, as was his wont, in grandiose terms about the NYSPCC: "There is a Providence in the affairs of men. . . . The slaves were the first freed from bondage; next came the emancipation of the brute creation, and the next the emancipation of the little children was about to take place."[6]

About two and a half weeks after Alexandre Liautard recorded the first case of horse flu in New York City, the Great Epizootic finally showed signs of subsiding. In a *Herald* article titled "Febrequo-bronchiatis," in which the paper tossed about a series of names for the Great Epizootic, they noted that "the interval of gloom and of paralysis that has affected the city . . . [after] the first attack of hippohinorrea was yesterday almost passed off and the mysterious mist of malaria seemed entirely lifted from our poor equine brutes." There were still plenty of sick horses, the story noted, but many had recovered, and the ill, on the whole, seemed on track to as well. But if stable owners were not vigilant in caring for victims not yet fully recovered, the *Herald* warned, the ailing animals could take a turn for the worse.

One sign that the epidemic was subsiding was that most streetcar lines were now employing the majority of their horses. Those horses, the *Herald* reported, "did not look either very emaciated or very weak." But there were exceptions. All was not well on the Bleecker Street line, where horses appeared as "sorry beasts that seem as if about to fall into pieces or melt away into melancholy grease spots."

Commerce was recovering and the downtown business district was coming back to life. Suddenly, the cartmen stands were home to healthy horses, able and willing to move goods, construction materials, and more from one place to another in the city. "Merchants had no difficulty in securing all the means they required for shipping their goods as soon as they were packed," reported the *Times*, and shipping rates had largely returned to normal. No surprise then that the "faces of the millionaires and merchants and bankers . . . [were] cheerful and confident." As the horses who powered transport and commerce began recovering, so too did those that pulled fire carts, hearses, mailcarts, and more.

Then, just as quickly as things got better, there was fear they might take a turn for the worse again, when cases of dropsy—what today we call edema or swelling from fluid in body tissues—were seen in horses in some of the stables around the city. "There has appeared in the large stables of the city a second phase of the horse disease," the *Herald* reported, "a relapse tending to dropsy which in several instances has proved fatal." The danger of dropsy was, in part, a function of where the swelling occurred. Affected horses, as a general rule, the *Sun* explained, looked "dull and listless," but their very life was at risk "[if] the swelling extends to the breast in the region of the heart and lungs."

There was general agreement among veterinarians that the cause of this new horse malady was in fact the long hand of the Great Epizootic: "It seems to affect those animals that were compelled to do work in the first stages of the epizooty," read one an article in the *Herald*. The Sanitary Committee of the New York City Board of Health concurred, writing in scientific bureaucratese that "the diseases spoken of as dropsy and glanders are sequelae or results of the former affection." Henry Bergh, who, after praising those stable managers who took good care of their animals during the epizootic, lambasted those who didn't: "Most of the horse owners, particularly the streetcar and stage lines, were too anxious to make money and drove their animals off their feet. Those that were allowed to recover only partially are now the ones that are falling victim to the new phase of the disease."

On November 15, the *Herald* reported there was "no perceptible interruption to travel" and the number of horses showing signs of dropsy was small compared to those that contracted horse flu—only a dozen of the 2,000 horses at the stables of the Third Avenue streetcar line were showing symptoms, and similar percentages were seen in other large stables. Still, the citizens had seen what it was like to lose horse power and were spooked. "The last phase of the horse disease is beginning to alarm owners," the *Herald* reported three days later, on November 18, "even more than did the epizooty weeks since. The symptoms are more deadly, and it is feared that 50% of the horses attacked will finally succumb under its ravages."

The alarmed response to the outbreak of dropsy, though understandable in light of the damage the horse flu itself caused, turns out to have been an overreaction. Within a week, reports of dropsy plummeted, and although estimates on the number of causalities and the rate of mortality are hard to gauge, they were likely very low, given how quickly dropsy fell from the pages of the newspapers. Less than a week after the initial dropsy scare, the *Herald* calmed readers by reporting that "the business markets today presented a scene of great activity and buoyancy."

The sense was that New York City's horses were at last safe and the city's main power source was back and that both the Great Epizootic and the secondary case of dropsy were now history. "All traces of epizootic have disappeared from sales stables," a *Tribune* piece reported on November 26, "and [if] we have mild and fair weather the prospects for an act of horse trade in the course of a week just commenced may be regarded as favorable."

Everyone, including veterinarians, agreed that whatever caused the Great Epizootic in horses had vanished by the end of November. Before it did, Adoniram Brown Judson, assistant sanitary inspector for New York City, estimates that it claimed the lives of about one thousand four hundred horses, or about 3.7 percent of the thirty-eight thousand horses in New York City. As horse flu had burned hot, there had been occasional "alarming stories . . . going the rounds of the city" that the disease afflicting the horses could be spread from horse to human, but there's no evidence at all that such tales were true, let alone evidence for cases of human mortality as a result.

Starting around October 21, it had been a very difficult five weeks for the horses and the citizens of New York City. As for the latter, New Yorkers had come to see just how reliant they were on their biotic mode of transport for almost everything that made their city great and prosperous. During that five-week spell, New Yorkers were not the only ones learning that lesson: at the same time that the Great Epizootic broke out in New York City, it also began its reign of terror in other great American metropolises including Boston and Chicago.[7]

# 7
# A Conflagration Without Horse Power

Upon my sole I've walked so much
Since horses have been hoarse
I feel like some pedestrian sport
In training on the course . . .

No car' os now will carry us
Urged on by driver's whacks
Oh for the wings of Icarus had
Although they went to wax

Then let us hope that horses may
Be freed from present woes,
And running on their feet again
Instead of at their nose.

—Lament of a human sufferer, *Boston Commercial Bulletin*

If the newspapers and citizens of New York City seemed unconcerned about the horse flu in the weeks leading up to the outbreak in their city, Bostonians and Boston's press were only slightly more apprehensive as the disease closed in on them. A short *Boston Globe* piece appeared on October 5, 1872 (and was reprinted in even shorter form on October 7), noting that "there is an epidemic . . . among the horses in Toronto . . . 700 animals are now sick with it." Readers of the *Boston Evening Transcript* would have read something similar that day, but with the addition that "almost all the horses belonging to the street-car companies and livery stables are afflicted." Three days later, in a short article titled "SERIOUS EFFECTS OF HORSE DISEASE," the *Evening Transcript* noted, "There is scarcely a horse in the city [Toronto] fit for work. The interruption to business is becoming very serious, as merchants are unable to ship or get goods from the depots. There are no hacks on the stand." On October 9, the *Globe* ran one sentence on the horse disease: "Toronto horses are suffering from catarrh." And then, as in New York City, the press became silent on the matter until ten days later, when on October 18 both the *Globe* and the *Evening Transcript* updated readers of the situation in Canada. Only on October 21 was the spread of the disease to the United States mentioned in Boston news, when the *Globe* wrote of cases of "the horse disease" in Buffalo.

Not surprisingly, given the scant newspaper coverage of the spread of the epidemic, between October 5 and when the disease was first recorded in Boston on October 22, the rhythms of the city remained unchanged. People went about their normal daily routines, walking or taking an omnibus, or more likely still, a horse car, to work or to shop or to enjoy one of the many leisure activities that the city had to offer. The Jordan Marsh Department Store at the corner of Washington and Avon Street was gearing up for the holiday season, advertising their "unusually elegant and elaborate assortment of Indian shawls of every choice variety. . . . Cashmere shawls . . . black silks . . . stripe silks and brocade silks" and more. W. H. Spooner's florist shop had Dutch bulbs, hyacinths, tulips, and ivies aplenty, and Palmer, Jacobs and Company, a hop, skip, and a jump from Jordan Marsh, housed a new assortment of new wool blankets to prepare for the harsh winters of Boston. There was also was the brand new Museum of Fine Arts for aesthetes, and indeed everyone, to enjoy.

As far as entertainment, just like in New York City there were a plethora of theaters in Boston where one could catch a show. In mid-October 1872, the Globe Theatre was hosting Lydia Thompson's Burlesque Show; the Boston Theatre was presenting a romantic play called *Rosedale: Or, the Rifle Ball*; the St. James Theatre was staging *Shin Fane*; at Tremont Temple, General Tom Thumb, just back from a three-year tour around the world, was regaling audiences with "Fascinating Performances and Laughable Sketches"; and *Othello* was being performed on the stage at the Boston Museum.

If they were so inclined, Bostonians could hop on streetcars, most carrying about thirty passengers, and head over to watch a baseball game at the South End Grounds on Walpole Street. Bostonians seeking literary enrichment could attend a lecture on the works of Scottish Poet Robert Burns at the Music Hall. Those who preferred a combination lecture center, art museum, and library could head over to the Athenaeum, which was hosting its *Forty-Ninth Exhibition of Paintings and Statuary*.

As was the case for the initial spread of the disease in Canada, and then to New York City, it is not possible to know for sure the exact route that the horse flu took from Markham or thereabouts to Boston: indeed, there may have been numerous paths. Still, if we accept a Buffalo-Rochester-Syracuse-Albany-New York City route as a reasonable starting point for how the epidemic reached New York City, and given that we know when the disease hit those cities, some conjectures on the horse flu's route to Boston become more feasible than others. In principle, given how much traffic there was between New York City and Boston, it is certainly possible that one path to the latter was through the former. But that's a rather serpentine path since it involves the disease heading 150 miles south from

Albany and then about 200 miles northeast to Boston. What's more, the earliest cases of the Great Epizootic in Boston were recorded at nearly the same time as the earliest cases were noted in New York City, making a New York City–Boston route unlikely. Given that we know that cases of the disease were recorded in Keene, New Hampshire, almost exactly halfway between Albany and Boston, shortly before cases were recorded in Boston, a more likely route to Boston would have been for the horse flu to have traveled nearly due east from Albany for about 185 miles.[1]

On October 22, the *Evening Transcript* alerted Bostonians of "several hundred horses prostrated," the result of "the horse disease which first made its appearance in Toronto . . . it appears to be . . . of an epizootic nature." When the disease struck at the Warner & Richardson Stable on Pembroke Street, about a mile southwest of Boston Common, it seemed to come out of nowhere and "took hold of a very large number of horses simultaneously." This was, the *Evening Transcript* informed its readers, "a practical illustration of the old saying 'misery loves company.'"

That same day, the *Boston Globe* assessed the number of horse flu cases as much higher: "Its speed has been so rapid that at present there are over two thousand cases, more or less severe," the paper reported. "The greatest sufferers are the express and horse railway companies most of which have some of their animals affected." But it was in the *Evening Transcript* story of October 24 that readers would find a more animated description of the possible effects of the Great Epizootic. "The place of the horse in civilization is likely to be rather impressively and painfully illustrated," the *Evening Transcript* piece began, "[for] an epidemic among the equine race is next in peril to an epidemic among the humans." The editors of the paper pleaded with their readers to "imagine the streets stripped of cars, carriages, carts and vehicles of all kinds, transportation—except by water or on railroads—at an end. . . . How great the discomfort, how serious the loss beyond easy calculation!" This epidemic, the *Evening Transcript* story proclaimed in no uncertain terms, could crush the city if unchecked: "The something in the air, or whatever else the cause may be that is turning stables into hospitals and carrying sickness and death among the most needful of dumb beasts is threatening no ordinary calamity." Instead, it would produce "a stagnation of business, disappointment of pleasure seeking and a derangement of affairs of severe detriment in a thousand directions."[2]

Boston residents were getting worried, and worry tends to spur rumor. Boston had dealt with a minor outbreak of smallpox in 1872, with Boston Hospital using a new special ward to isolate the 220 infected patients, and when the Great Epizootic was in the city, the *Globe* fanned fears of other maladies on the horizon that were tied, the paper claimed, to what was happening to the city's horses. That same year in Hungary, in the cities of Buda and Pest, a major outbreak of cholera had struck, and the *Globe* warned readers that "physicians tell us that they fear this horse distemper is but the *avant-courrier* [advance guard] of that terrible scourge, the cholera."

On October 25, the *Globe* informed its readers that the spread of the disease among horses "has been so rapid that it can now, without exaggeration, be regarded as a great public misfortune. The traveling public is particularly incommoded, owing to the irregularity or the total stoppage of the horse cars and coaches." Some papers were nicknaming the epidemic the "exercise promoter," the "boot-maker's friend," and "The Corn Doctor's Blessing." By the next day, Boston appeared to some residents to be little more than a horse ghost town. "The fears of a suspension of local travel and traffic on account of the malady afflicting horse flesh generally have become realities," the *Evening Transcript* reported. "Pedestrians from Cambridge, Charlestown and South Boston . . . were struck with the almost complete desertion of the streets, while the sidewalks were positively crowded with umbrella covered humanity striving

**Figure 7.1**
A cartoon during the Great Epizootic. From *Harper's Weekly*, November 16, 1872.

to reach the scenes of daily toil at the accustomed hours. . . . Where hundreds of express wagons were to be seen in the court square and other rendezvous yesterday, scarcely a dozen could be counted at the corresponding hour today" (figure 7.1).

Miasma again was the presumed culprit, the cause being, at least according to the *Evening Transcript*, "atmospheric influence." Talk of miasma would not have been completely foreign to readers of Boston's newspapers, since the *Evening Transcript* the prior fall had published that "sunlight dispels more than one kind of darkness, and true pure air disinfects the soul chambers, as well as those where lurk miasma [and] disease." Still, when presented to the general public, news coverage comparing the miasma and germ theories could be very confusing. In one story, run during the early days of the Great Epizootic, the *Boston Globe* told its readers that it was "an important point to discover . . . whether the disease as seen in this city, results from an inhalation of germs floating in the atmosphere or whether it owes its origins to direct contact of animal with animal." Here, readers could be excused if they were not certain whether the phrase "inhalation of germs floating in the atmosphere" used the word "germ" in the very generic sense of "that which causes diseases," as in the chemicals in putrid air and water that miasma theory considered to be the causative agents; or whether the *Evening Transcript* was suggesting that microorganisms were floating in the air and that when they were inhaled they caused disease, but that the disease was not contagious. Readers' confusion was inevitable because the newspaper made no attempt to clarify which of these possibilities it was implying.[3]

As in New York City, during the Great Epizootic both humans and oxen were sometimes employed in lieu of horse power in Boston. Jordan Marsh Department Store delivered their goods using "pull-men" wagons, a play on Pullman railway cars, drawn by nine men, and other companies did likewise. The *Evening Transcript* could not pass up a chance at taking humorous swipes at this turn of events, applauding the pull-men's "equine-imity of temper" as "many men have of late most commendably made beasts of themselves . . . it is in no spirit of horse-stillity that we say, 'keep your horse still. If this be not philanthropy, it is phil-hossophy."

It wasn't just goods that could not be transported in the absence of horse power. Hearses to transport the beloved to their eternal resting places were next to impossible to find. "One of the most melancholy effects of the epidemic among the horses in Boston results from a difficulty which undertakers experience in obtaining horses for funerals," one story read. "In one instance a coffin containing a corpse was conveyed to the streets in a hand cart."

Streetcars had all but stopped running. The Metropolitan Railway reported that nearly half of their 1,200 horses were down with the flu. The situation was even grimmer at the Union Railway where 550 of the 700 horses were too sick to pull a streetcar. No numbers were given for the horses at the South Boston Railway, but the *Globe* did note that they had "the usual proportion of diseased [horses]," which given the numbers coming from the Metropolitan

**Figure 7.2**
Men pulling a streetcar in Boston during the Great Epizootic. From *Frank Leslie's Illustrated Newspaper*, November 16, 1872.

and South Boston lines was hardly a good sign. As for omnibus travel, the *Globe* is brief and to the point, using one line as an ominous representative for the condition of all: "The omnibuses of the Citizen's line have not yet made their appearance in the streets, all of the horses employed being considered unfit for work." The only positive note to report was that a single steam-powered streetcar dummy engine was being tested between Cambridge and West Boston, and early trials were promising. *Frank Leslie's Illustrated Newspaper* captured the multifarious effects of the epizootic in Boston in an article it published that included five striking drawings: a streetcar pulled by people, a streetcar drawn by oxen, a bursting-to-the-brim overcrowded streetcar, a milkman carrying massive milk churns in lieu of horse power, and grocers pulling wagons of produce to their customers (figure 7.2).

"In future years, when the children and grandchildren of the present generation are told of the strange visitation which Boston received in the year 1872 and the extraordinary scenes and circumstances connected therewith," the *Globe* philosophized, "they will be hardly able to imagine the queer predicament in which we are placed. No horse-cars, no express wagons, no hacks, no coaches, nothing drawn by that most useful and docile animal, the horse." On a lighter note, but also one that reminds readers of the perils of relying on living creatures as their source of power, the *Globe* opined that "the iron horses are the only ones now exempt from the epizootic malady."

Horses working for the city were hit just as hard as their private-sector counterparts. Fire department horses had an especially hard life in general in Boston, because the city, in an attempt to replace hand-cranked pressure to shoot water through hoses, had purchased twenty-one steam-powered pressure systems mounted onto fire engines, many of those from the Amoskeag Locomotor Works Company of Manchester, New Hampshire, fifty miles northwest of Boston. These systems allowed water to be sprayed farther and faster, free from the fatigue of the firefighter turning cranks. There was a sense of wonder and awe that these machines captured, so much so that some considered them worthy of verse, as in a poem titled "The Steam Fire Engine."

Behold! How she shines in her beauty,
Resplendent in silver and gold;
Ne'er shrinking from doing her duty,
When worked by her members so bold;
. . .
"She saves us a deal of hard labor,
Her muscles seem never to tire—
Admired by stranger and neighbor,
Whene'er she's at work at a fire.
"The Steamer! We'll ever adore you;
In praising you we never tire;
Hand-engines were nothing before you,
Nor compared with you when at a fire."

But these steam-pressured systems were also very heavy, making the entire fire engine a behemoth for horses to pull. And now, on top of that, John Damrell, chief of the Boston Fire Department, reported that two days after the Great Epizootic struck Boston, two-thirds of the fire department horses were down with the flu. As a result, he instructed drivers of fire carts that as soon as their horses showed signs of the disease, they should be brought to the stable, blanketed, and fed mash bran. Damrell "contemplate[d] that should the disease spread further among them, [he would] request . . . citizens to 'man the rope,' as in yore, when an alarm of fire is given." Two days later he ordered the board of engineers "to obtain volunteers sufficient to double the number of men connected with the engine, hose and hook and ladder companies." One glimmer of hope on the horizon was that there was a rumor afloat that the fire company might at last be getting a very expensive new Amoskeag self-propelled fire engine that used steam not only to create pressure for water to douse fires, but also to power the movement of the fire engine itself: no horses required.[4]

As the Great Epizootic unfolded, and horses everywhere were coming down with the flu, Bostonians were learning of Henry Bergh and the ASPCA. In his October 8, 1872 lecture titled "Masters of the Situation," Boston publisher and editor James Field—who counted Longfellow, Dickens, Hawthorne, and Thackeray as friends—spoke glowingly of Bergh and his organization. "All honor to Henry Bergh of New York," Fields told his audience, "the firm and unselfish advocate for that part of creation which cannot ask kind treatment for itself; the man who has spoken so effectively for those poor dumb mouths that have so long silently pleaded for protection from injury at the hands of man. . . . It is a sacred mission to which this man has been called and among the world's benefactors he has proved himself a noble master of the most difficult situation."

Bostonians soon learned of Bergh and the ASPCA's actions during the Great Epizootic in New York City. "Brother Henry Bergh, at once the guardian angel and the patron saint of the bestial race, exercised functions of his high office yesterday evening with effect," read a piece in the *Globe*. "He gathered his tribes together and took the warpath. Standing at the junction of [New York City's] Third and Fourth Avenues, at the Cooper Institute, he displayed his professional shield, and, in the name of the august Society for the Prevention of Cruelty to Animals, ordered the brutes to stop driving the gasping beasts. There was of course a rumpus; but Bergh, being reinforced by the police, carried the day."

If Bergh's ASPCA was the ur-society for the prevention of cruelty to animals, during the Great Epizootic, on a day-to-day basis, it was George Angell and his Massachusetts Society for the Prevention of Cruelty to Animals (MSPCA) that Bostonians would see out on the streets, keeping a close eye on how horses, especially those on streetcars and omnibuses, were being treated. Bergh instructed his Massachusetts counterpart to "make arrests yourself & go into court & prosecute, as I do 3 times per week. Make 'your name,' as mine has become, a terror, for good doing." Angell would use his agents to prosecute when absolutely necessary, but even though he was a lawyer, he was much less litigious than Bergh and preferred education to prosecution. On that front, he published *Our Dumb Animals*, the first journal devoted to matters of animal protection. Pitched largely at young readers, during the first year alone half a million copies of *Our Dumb Animals* were published, and Angell convinced the police to help him distribute the journal into classrooms all over Massachusetts. Students across the Commonwealth were soon signing "The Humane Pledge," which read: "I will try to be kind to all harmless living creatures, and try to protect them from cruel usage."

Angell's approach to animal cruelty that emphasized education over prosecution was also adopted in other ASPCA spinoff societies. In Philadelphia, Caroline Earle White, whose family included many abolitionists and advocates of women's rights, took on combating animal cruelty as her moral mission. Inspired early on by the works of Jane and Ann Taylor, which instilled empathy for animals in the minds of young children, White visited Bergh in 1866 and worked with him to create the Pennsylvania Society for the Prevention of Cruelty to Animals (PSPCA).

As was the case far too often when it came to far-sighted woman reformers, White learned that in the patriarchal world of the early Gilded Age, organizing was one thing, but being part of the power structure of what was organized was another. The PSPCA thanked White for her efforts but saw no place for her and suggested she create a "woman's branch" of the PSPCA, which she promptly did. It was there that White adopted Angell's education-first approach. White also formed children's branches of the PSPCA, where young officers were adorned with a horse badge and ornate sash. She distributed copies of *Our Dumb Animals* far and wide, so that children would understand that when they acted to prevent cruelty to animals, they were acting as responsible citizens. "When the human being looks at the sufferings of the inferior animal," one issue of *Our Dumb Animals* preached, "and with a heart full of love seeks to alleviate those sufferings, from no selfish consideration whatever, but because it is suffering, he is patterning after the true conception of . . . boundless munificence and love and philanthropy."

It is hard to assess the impact of Angell's or White's education outreach on happenings during the Great Epizootic, but it seems likely that Angell's program played some role in explaining a curious difference between those who owned horses, be it many or few, in New York City versus Boston. Bergh and his agents faced daily resistance when horse flu swept into New York City, but the situation was far less tense in Boston. Over and over during the epidemic, Frank Eay, Secretary of the MSPCA, used the Boston newspapers to update readers on his society's actions and reassured them that, although there were exceptions, in general "we have found our people, corporations and individuals, willing and anxious to avoid working their horses. . . . In a large number of cases, our agents have directed that horses badly diseased should be put up [in stables] and owners have readily complied." There's insufficient data to know if that more cooperative response led to reduced mortality or quicker convalescence, but it may have.

This is not to say there was no resentment toward the MSPAC on the part of streetcar companies. "The directors of the Metropolitan Railroad regret to be obliged to apologize to the public for the cessation of the trips of their cars yesterday," read an article that Joseph Draper, president of the that streetcar company, wrote for the *Evening Transcript*. "But the agent of the society for the Prevention of Cruelty to Animals gave them notice . . . that they must put up all their horses . . . we therefore for the day . . . withdrew our cars." But that withdrawal was under protest, Draper felt obliged to make clear: "We deem it right for our patrons to know our position. We were incorporated to carry the people. Families have established themselves on our lines at a distance from their places of business or employment, depending upon the cars to carry them. . . . We venture to say that more persons will become sick, and, perhaps die, from exposure in being compelled to walk home . . . than there are horses who would have been in any degree injured."

As November rolled in, the horse flu was still causing havoc in Boston. Most streetcars and omnibuses were running partial service, at best, and the *Globe* informed readers on the first of the month that "business continues to suffer from the epizootic among horses, and to a much more serious extent than was at first supposed." In a rather somber tone, the paper added: "Meanwhile the community must be patient and hope for the speedy recovery of the equine species, upon whose health the affairs of business seem to depend so largely."[5]

Though many fire department horses were down with the flu, Boston was fortunate in that it possessed a more sophisticated and accessible fire detection system than did New York City. Back in

1852, forty fire alarm boxes had been placed across the city, the brainchild of physician and inventor William F. Channing. These boxes employed telegraph technology, such that activation, by cranking a handle when a fire was detected, triggered a telegram that included coded information on the location of the fire box to a fire alarm office at the City Building (Old City Hall today) on School Street. The operator who received the information immediately sent another telegram, via a different set of telegraph wires, alerting the appropriate fire house of the danger. It was not long before the crank handle was replaced in newer "sector boxes" with a handle that one simply pulled down to send the telegram. When the Great Epizootic struck, the streets of Boston were home to more than one hundred and fifty of these fire alarm boxes.

At 7:24 p.m. on November 9, 1872, as the Great Epizootic continued, though not as furiously as it had early on, the alarm at Fire Box 52 went off. A fire was blazing in a five-story, 5,000-square-foot building at 83–85 Summer Street, on the corner of Summer and Lincoln Streets. That building housed flammables galore, being home to Tebbitts, Baldwin, and Davis Dry-Goods, as well as Damon, Temple, and Co., dealer in fabrics and clothing, and Alexander K. Young and Co., producer of ladies' hoopskirts. The second alarm was sent at 7:29 and subsequent alarms at 7:34, 7:45, and 8:00 p.m. Soon, additional alarms were transmitted from Fire Boxes 48 and 123. The Great Boston Fire of 1872 had begun. "Currents and counter-currents were driving the flames in every direction . . . [the fire] swept through our streets with the power of a tornado," Chief Damrell wrote. "The roar of this Niagara of destruction was appalling . . . gas and air explosions in quick succession, falling walls, and intense heat, made it a terrible experience to that faithful organization which it was my privilege to command."

Boston's firefighters, including those who attempted to douse the flames from the tugboat *Louis Osborn*, put up a valiant fight, joined

**Figure 7.3**
A fire engine with a steam-powered pressure system at a scene of devastation after the Great Boston Fire of 1872. From Amoskeag Reserve Engine Company of New Hampshire, accessed October 2024, https://arecnh.com/projects/.

by firefighters (and engines) from twenty nearby towns and villages who responded to urgent telegrams from Damrell requesting aid from any company within a vicinity of fifty miles. But sixteen or so hours after the fire began, when it was at last put under control at the corner of Washington and Milk Streets, 776 buildings, many of them warehouses and businesses, were laid waste, a thousand residents were homeless, and estimates of the damage hovered around seventy million dollars (roughly $1.5 billion today). Nine firefighters (including volunteers) lost their lives in the line of duty that night. Two more firefighters died later from injuries. Civilian casualties were remarkably light: on the order of two dozen (figure 7.3).

As might be expected for a conflagration of this magnitude, there were many factors at play as to why what might have just been another of the 640 fires in Boston in 1872 instead became the Great Boston Fire. Time was certainly one. The fire was spotted first by two police officers, who saw it from Prison Point Bridge at 7:08 p.m., but for reasons that were never fully explained, the first alarm was not sounded for sixteen minutes. The elevator shaft where the fire began was another aggravating factor. The cause of the fire was never determined, but once it began it shot to the roof quickly because the elevator shaft was made of wood and had no escape hatches, allowing the fire an easy path upward.

Though some buildings were made of brick or stone, window frames and roofs were constructed from wood. The mansard roofs sitting high atop the buildings adjacent to 83–85 Summer Street were a disaster waiting to happen. Built from pine and tar, with steep sides and a double pitch, these roofs made excellent kindling for the fire once it began. Fire Chief Damrell had complained many times that these roofs were dangerous, but they were quintessential Boston architectural design at the time, and his complaints led nowhere. Damrell was not the only one concerned. Shortly before the fire, insurers from both Liverpool and London had told the Massachusetts insurance commissioner they were considering cancelling insurance policies for a chunk of the city because of concerns about the wood used in construction of many of the buildings that eventually went up in flames.

Once the fire began, the narrow roads in the neighborhood made it hard for fire engines to navigate. There were also problems with the supply of water. No underground reservoir was near where the fire broke out. Hydrants were hundreds of feet apart, some had the wrong couplings for fire engines from the suburbs, most allowed connection to just a single hose, and others were rusty, impeding water flow. Damrell had been worrying about all these matters for years, but there was little he could do. There was also the matter of allowing citizens to use gunpowder to blow up certain buildings to slow the spread of the fire. At first, Damrell would not permit this, but he relented after pressure from a "citizens committee" that he met with in the mayor's office as the blaze worsened. Not long after the Great Boston Fire, however, Damrell wrote that "the use of gunpowder in this district was contrary to my judgment," and years later he added, "I have never forgiven myself for the mistake made in granting authority of the kind."

The flammable architecture and construction of buildings, the urban planning that created narrow roads, the lack of foresight with respect to the water supply, and more made Boston a firebomb ready to explode. But the fact remains that once the blaze began, the lack of horse power to pull the fire engines into place quickly, the result of horses down with the flu, was a significant factor in the initial rapid spread of the inferno. Even the extra men put on duty during the Great Epizootic simply could not get those fire engines there quickly enough, no matter how hard they pulled. Louis Sullivan, who would come to be known as "the father of skyscrapers," described the absence of horse-powered fire engines on the blaze. In his autobiography, *The Life of an Idea*, Sullivan, who writes in the third person, recalls that he had been out on the streets of Boston near where the fire broke out on the evening of November 9. "Louis saw this terror from its trifling beginning—a small flame curling from the wooden cornice of a building on the north side of Summer Street," he wrote. "The street was night-still. It was early. No fire engine came. Horses were sick, 'epizootic' was raging. Engines must be drawn by hand . . . a few more people gathered, but no engine came. . . . glass crackled and crashed, flames burst forth madly from all windows, and the . . . dark flames behind them soared high, casting multitudes of sparks and embers abroad, as they cracked and wheezed . . . [the] roof fell, the floors collapsed. A hand-drawn engine came, but too late."

Sullivan's first-hand account was just one of hundreds that, along with data provided by the Boston Fire Department and other city agencies, was folded into a report by the commissioners appointed to investigate how and why the devastating fire started and quickly became uncontrollable. The report makes clear that while a few of the first engines to arrive were horse-powered, that was the exception. The rumor that the Boston Fire Department might be getting an Amoskeag self-propelled fire engine turned out to be just that: a rumor. And because fire engines were pulled by people instead of horses, the ability to fully combat the blaze "was delayed from three or five minutes for the nearest, to forty minutes for the most distant," noted the *Report of the Commissioners Appointed to Investigate the Cause and Management of the Great Fire in Boston.* "No one can tell what was the cost of these minutes to our city."

The *Report of the Commissioners* determined that "the horses of the department were not used, because they were generally unfit to be used." Those sick horses, they decided, would not have been able to pull a fire wagon faster than people could, and that "they were rightly left in the stable or the hospital." On that count, Fire Chief Damrell and the Boston Fire Department were hardly at fault, but that was just part of what the commission had to say about the effect of the Great Epizootic on the Great Boston Fire. They, like everyone else in Boston, were aware that by the night of November 9, the epidemic had waned a bit and there were, in fact, sufficient horses scattered across the city that were healthy and that might have been commandeered in the days before the fire to prepare for a worst-case scenario.

In fact, Damrell, on October 26, had issued an order instructing that "the drivers of horses which are incapacitated for work are authorized to seize, in case of fire, any horses they may find upon the street." The problem, as far as the commission was concerned, was that the order should have been acted on preemptively, not just in response to a fire, so that the department would be ready when a worst-case fire occurred—as it did fifteen days later. Horses, the report read, "were in general use in the city, and before this time might readily have been obtained for the use of the Fire Department. There is no evidence that this occurred to anyone, but it ought to have occurred to those whose special duty it was to guard the community from the perils of fire." Years later, Damrell disputed the claim that "this occurred to no one," and wrote that he put together a subcommittee to visit all the stables in the city to obtain horses to use in his fleet—but none were found. In any case, no such horses were procured, and the commission left no doubt as to what the consequences of that mistake were. Immediately following their castigation of Damrell and the Fire Department leadership for want of preparation, they noted, "We cannot but feel that the earlier arrival of aid, between 7.30 and 8 P.M., might perhaps have prevented the flames from crossing from Summer [Street]. . . . This was the 'key of the fire' at its outset; and when it was lost, the flames became almost uncontrollable."[6]

In the days following the fire, Boston newspapers reported almost exclusively on the conflagration, though the occasional piece could be found on the epizootic. On November 13, there was a five-line story noting that while "the horse disease is rapidly disappearing," a new complaint of dropsy was being registered by owners whose animals "were longest afflicted with the distemper." The next day, the MSPCA published an article in the *Boston Globe,* going to pains to let the public know that no matter how much the MSPCA members valued animal life, they valued human life more. "In an emergency like that of Saturday night no objection of ours would have prevented the use of the department horses," wrote Frank Fey, the MSPCA's secretary, "even at the sacrifice of their lives."

By the third week in November, the Great Epizootic had done its damage in Boston and was hardly mentioned again in the press, except in the occasional black humor piece, as when the *Globe* wrote in its "Minor Items" column that one name for the epidemic that had swept through was Hipporhinephlegmatoblennocalastalagamus. "There is a mistake somewhere in that name," the paper added, "and an awful suspicion is afloat that its designer has omitted more than one letter of the alphabet." Black humor aside, as Bostonians tried to process that a good portion of their beloved city now lay in ashes, in part because of a lack of healthy horses, they understood all too well that the Great Epizootic had been anything but a minor item.[7]

# 8
# Chicago Never Does Things by Halves

Smiting the lowly, nor sparing the high . . .
Is it Never to cease till the horses all die?
What shall we do for it? What is the best?
Cider, warm blankets, or plenty of rest?
Low is the head that so proudly was rearing and drooping the neck that was curved in his pride.
Dimmed the fierce eye, and deadened the hearing, faint throbs the once wildly pulsed purple tide.

—"The Epizootic," a poem by an "Excited Horse-Fancier," *Chicago Tribune*, November 17, 1872

"Chicago never does things by halves," read an article in Chicago's *Tribune* on November 5, 1872. "We mean to be champions in everything, even in the epizootic line. This city will be as much superior to other cities in that disease as it is to-day ahead of them on conflagrations." The conflagration reference is, of course, to Chicago's own devastating fire, the most famous in American history, which tore through the city between October 8 and 10, 1871, a year before both the Great Epizootic hit Chicago and before Boston went up in flames.

The *Tribune* painted a pretty picture of the city a week before the Great Chicago Fire: "All was animation and prosperity in the Chicago of the 1st week of October, 1871." The city had "waxed fat and was proud, exceedingly proud of what it had done and what it was to do, of its merchant princes," the paper wrote, "its tunnels under lake and river . . . its projected parks, its contemplated railroads, its proposed new buildings and its great and certain future." Then, nine days later, one third of the city was in ashes: downtown Chicago needed to be rebuilt, almost from scratch. When the Great Epizootic struck, that rebuilding ground to a standstill.

Before the Great Chicago Fire broke out, more than 50,000 buildings stood in the city limits. Almost half were made of wood. On average, two fires a day broke out in the city, but after a fourteen-week drought, the summer of 1871 saw more fires than that average. In September, only three-quarters of an inch of rain was recorded for the entire month. The week prior to the Great Chicago Fire had seen twenty fires, including one that began at about 10:00 p.m. on October 7 at the Lull-Holmes Planing Mill at 209 South Canal Street. This blaze quickly spread to Jackson, Adams, and Clinton Streets before it was extinguished, exhausting Chicago's 185 firefighters and the fire horses that powered the seventeen steam engines in the city fleet.

On October 8, at approximately 9:00 p.m., a new fire broke out in Patrick and Catherine O'Leary's barn at 137 Koven Street, which lay just a few blocks west of the south branch of the Chicago River, and just eight blocks south of where the Lull-Holmes fire had burned the prior night. At some point, next-door neighbor Daniel Sullivan roused the O'Learys from their sleep and together they began to save what they could. Exactly what caused the blaze will likely remain a mystery, though many still accuse Mrs. O'Leary's poor cow of being an accidental arsonist, a rumor first printed by the *New York Tribune* on October 11, 1872.

A watchman on duty at a courthouse tower alerted the fire department of the blaze at about 9:30 p.m. Soon, a firefighter at the America Fire Hose Cart Company at 31 Blue Island Avenue saw the flames for himself and he and the company's horseman got the horses ready and drove the *America* to the fire, where they hooked their hose to a fire plug and began fighting the blaze. "Little Giant" Company No. 6 on Maxwell Street arrived soon after. Within an hour, seven fire companies were on the scene, but the fire, powered by strong winds, was already out of control. By this point, a fire tornado had carried burning wood to St. Paul's Church at Clinton and Maether Streets, five blocks north of the O'Leary barn. Soon the church was burning so hot that firefighters on horse-drawn fire carts could not get close enough to do anything but watch as the church steeple went up in flames. Next, the fire crossed 400 feet to the east of the Chicago River—parts of which itself were ablaze from the fire-ignited garbage and oil on the water's surface—and burning debris set alight the home base of Parmelee's Omnibus Company, the Chicago Gas Light and Coke Company, and sections of a large, run-down, crime-infested neighborhood known as Conley's Patch.

By midnight, Chicago Mayor Roswell Mason was sending panicked telegrams to nearby cities to send help, and as Fire Chief Damrell would do in Boston, he finally acquiesced to pressure from local businessmen and gave his blessing to the use of gunpowder to create a firebreak. Firebreak or no, the situation by this point was already hopeless because the fires at Parmelee's Omnibus Company, the Chicago Gas Light and Coke Company, and Conley's Patch had merged into one giant inferno. The conflagration then moved north and engulfed Crosby's Opera House, which the prior summer had invested \$80,000 on a lavish remodeling. Two blocks west of Crosby's sat the city's main court, with a jail located in its basement. At 2:00 a.m., Police Captain Hickey ordered that less dangerous prisoners be released and more hardened criminals placed in handcuffs and escorted elsewhere. A half hour later, the fire had crossed another branch of the Chicago River and railway cars carrying kerosine at the Chicago and Northwestern Railway yards exploded.

At 3:20 a.m., after the inferno had moved further north and east, burning debris was blown to the roof of one of the city's main pumping stations, at Chicago Avenue and Pine Street. The wooden shingles on the roof had been replaced by slate, but underneath the slate was a tinderbox of wood that ignited. "The pumping works became an utter wreck," was how DeWitt Cregier, city engineer and future mayor of the city, described the scene. "Nothing but the naked walls of the building and the broken and blackened skeletons of three engines were left to mark the spot from whence only a few hours before flowed millions of gallons of pure water for the comfort and convenience of our citizens."

If there was one building in Chicago that was designed not to burn, it was the Post Office and Customs House at the corner of Dearborn and Monroe Streets, with its stunning exterior constructed of solid stone. Even the building's shutters, made of iron, were supposedly fireproof. Yet at 7:00 a.m. on October 9, the Post Office and Customs House too caught fire. Further north, William Ogden's massive lumber yard and the McCormick Reaper Works burned as well that morning. Later that day, former Civil War General Philipp Sheridan ordered the use of gunpowder as a firebreak as the blaze burned on Michigan Avenue. Finally, late that evening, the Great Chicago Fire began burning itself out, far north in the city, near Fullerton Avenue. Then a much-needed rain storm on the morning of October 10 put an end to the dwindling flames.

When the Great Chicago Fire finally burned out, three-and-a-half square miles (9 square km) or about a third of the city lay in smoldering ruins. More than seventeen thousand buildings were destroyed, causing approximately two hundred and twenty million dollars of damage (about five and a half billion dollars today). Somewhat miraculously given the scope of the fire, only somewhere between two hundred and three hundred people were killed. But an estimated one hundred thousand Chicagoans—nearly one third of the city's population—found themselves homeless.

On October 11, Mayor Mason established martial law in Chicago, to be administered under the rule of General Sheridan. "The preservation of the good order and peace of the city is hereby intrusted [sic] to Lieut. General P.H. Sheridan, U.S. Army," Mason decreed. "The Police will act in conjunction with the Lieut. General in the preservation of the peace and quiet of the city." By October 24, 1871, when it was clear the continued peace and quiet of the city was assured, martial law was lifted. Almost immediately, the rebuilding of the city commenced.

That rebuilding process was powered by Chicago's 25,000 horses. The first matter at hand was to haul away debris from the more than seventeen thousand buildings that were now charcoal and ash. Horse-powered vehicles of all forms were commandeered for that monumental task. The construction of new residences, businesses, warehouses, municipal buildings, bridges, viaducts, schools, churches, hotels, theaters, music halls, opera houses, libraries, and more also depended on horse power. From the time of the Great Chicago Fire up to the outbreak of the Great Epizootic, thousands of building permits were issued: the footprint of those buildings measured 43,413 feet of streetside facade and virtually all of the construction material was hauled by horses at one stage or another. "Hannibal crossing the Alps, and Napoleon retreating from Moscow," the *Lakeside Monthly* wrote, "were mere skirmishes in comparison with the battle with snow and ice waged all winter long by Chicago builders. . . . Truer heroism in the endurance of hardship was never shown." To be fair, the same accolades should have been showered on the city's horses. But they weren't.

Supplies, including food for 100,000 homeless Chicagoans and twice that number who were housed, needed to be moved within the city, as well as from the railroad depots into the city, and the vast majority of the deliveries relied on power supplied by horses. In the twelve months that followed the fire, the Chicago Relief and Aid Society alone distributed 1,926,460 pounds of flour, 467,514 pounds of pork, 556,350 pounds of beef, 635,925 pounds of bread, 151,655 pounds of crackers, 61,898 pounds of coffee, 259,744 pounds of sugar, 922 bushels of potatoes, 7,627 bushels of onions, 214,314 pounds of soap, and 98,356 pounds of candles. In just a single week after the conflagration, 10,737 mattresses, 25,339 blankets, 4,653 tons of coal, 4,459 stoves, 22,531 pairs of shoes, and 165,652 items of clothing were distributed across Chicago's Wards 1–6. The delivery chain that brought those millions of pounds of food and supplies to the people of Chicago was, at various links, almost always powered by horses.

On the one-year anniversary of the Great Chicago Fire, the *Tribune* ran a huge, nine-page story on the rebuilding of the city. "The wonderful activity of the citizens of Chicago, their energy and enterprise, like the fire and its incidents cannot be exaggerated," the paper proclaimed. "Never before has the world seen, and never again will it see, such a marvel of growth as the New Chicago of today can boastfully point to." That claim may have been a bit hyperbolic, but just a bit. Still, there was much to be done when, two weeks after that anniversary issue of the *Tribune* hit the stands, the Great Epizootic arrived.[1]

The Chicago Board of Health pegged "the first cases [of horse flu] occurring here . . . in a stable . . . in some horses brought from Canada." How those horses got to Chicago is unclear, but it may have been via Union Stockyard—a 475-acre market at Exchange and Halsted Streets. For all practical purposes, this stockyard was a giant auction house that largely dealt in cattle, hogs, and sheep. In 1872, nearly three quarters of a million head of cattle, more than three and a half million hogs, and nearly a quarter million sheep arrived and were sold at and transported via the yards. But horses too, were bought and sold there. That same year, 12,145 horses, many shipped to Chicago via the rails, were sold in the Union Stockyard. There is no hard evidence one way or another, but some of those horses, perhaps even some from Canada, may have been infected. Rumor had it so: ten horses that had arrived from Canada on October 20, it was said, brought the disease with them and the outbreak could be traced back to them. Of course, as with New York, Boston, and elsewhere, the disease may have entered Chicago via many routes simultaneously, including not just horses imported from Canada or those sold at the stockyard, but also via a stepping-stone trek westward from Detroit.

On October 23, 1872, the *Inter Ocean*, one of Chicago's many daily papers, informed its readers that "orders have been issued at several stables to admit no new horses except on positive assurance that they do not come from Canada or Western New York." There was more: "The directors of several street railway companies met today and agreed to petition the managers of the various steam railway lines to refuse to bring horses here from the infected districts." The *Tribune* reminded Chicagoans just how reliant their city was on horses, particularly in the face of the massive reconstruction still underway. "Forewarned is forearmed," they told their readers. "The veterinary surgeons and the owners of horses in this city cannot be on their guard too promptly. Chicago is more dependent upon horses today than any other city in the union. The literally immense tide of street travel and the necessity for the use of large numbers of horses in our building operation, in addition to the ordinary demand for purposes of personal pleasure and convenience are such that an interruption at the present time must necessarily work serious results. . . . An ounce of prevention is worth a pound of cure."

Just a day after the *Inter Ocean* article alerted the city's residents to the danger, they were already receiving mixed signals. The *Evening Post* warned readers of "the dangers the pestilential breath of that mysterious disease which has been . . . decimating the equine populations of Canada and which during the last month has made havoc in the stables of New York." That pestilential breath had now made its way into the stables of streetcar companies such as the West Division City Railway Company. The editors at the *Tribune*, as they had earlier, remained very worried, particularly after reporters interviewed a Dr. Tuthill, the veterinarian in charge of the 320 horses at the West Side Street Car Company. "Should the disease prove to be the dreaded epidemic, as [Tuthill] had no doubt it would," the *Tribune* reported, "the dingy, dirty, ill-ventilated, character of the livery and boarding stables, omnibus stables, and auction stables together with the crowding which they were compelled to put up with, would produce a terrible mortality." But, readers of the *Chicago Evening Mail* were being sent a very different message: "It is fortunate that horses can't read the morning papers," the *Evening Mail* noted, "otherwise our Chicago equines would be scared to death today. . . . However as there is no evidence of the presence of the dreaded epidemic in this city now, they would not thank the morning papers for their unnecessary attempt to be sensational." It was only a matter of a day or two more, though, before all the city's papers, including the *Evening Mail*, were reporting on just how precarious a position their city was in, after making so much progress on recovering from the fire a year ago.

Soon there was talk of "epizootic vapors" as the cause of the afflicted horses. Daily equine mortality counts were published in most city newspapers, often under morbid headlines like the *Evening Post*'s "Death Mounted on a Pale and Plague-Stricken Horse Makes his Appearance." City officials quickly took extraordinary measures to try and curb the spread of the epidemic. Though it is unclear whether they heeded the request, on October 29 the mayor and the sanitary superintendent asked steam railroad companies "to refuse to carry horses into or near Chicago during the prevalence of the epizootic." The Illinois Humane Society, which had recently changed its name from the Illinois Society for the Prevention of Cruelty to Animals, sent agents to the streets. In general, the society was pleased with the manner in which omnibus and streetcar companies were treating their horses. Newspapers ran stories of the Great Epizootic daily and were sending reporters to stables around the city to assess just how bad the situation had become. What they told their readers confirmed what they could see with their own eyes, as the number of streetcars and omnibuses that transported people to their daily routines was plummeting. A week after the first horse flu cases were reported, the West Division City Railway Company had dozens of horses down. At Parmelee's Omnibus Company, which had recovered from the damage of the Great Chicago Fire, more than 80 percent of the horses were sick.

On November 2, the *Tribune* was concerned enough about the Great Epizootic that they offered a $5,000 reward to anyone who came up with a cure. That same day, a headline in the *Inter Ocean* read, "The Horse Disease: Ten Thousand to Twelve Thousand Horses in Chicago Unfit for Use." The opening sentences of the story drove home the message: "The idea that the horse disease had done its worst in Chicago, that it would prove light and visit but a few stables, has proven erroneous. It went through the town yesterday like wildfire." The story continued: "Of the 20,000 to 25,000 horses in the city, at a moderate estimate, one half are on the sick list and the poor steeds that are yet in the streets indicate by their appearance that the culmination of disease has not yet been reached."

Streetcars in the city carried an average of 100,000 people a day. The *Inter Ocean* stated that "whether the residents who are dependent on streetcars and omnibus lines will succeed in reaching business at the usual hours this morning is more than doubtful." A *Tribune* story with the subhead "The Streets Deserted. . . . Pedestrianism the Most Fashionable Mode of Transport" observed that Chicagoans' "legs that seldom walked were called upon to transport their owner many miles during the day." The same article noted that a special meeting of the city council would examine the possibility of using steam dummies to help transport Chicagoans in lieu of horse power.

The city had access to several steam dummies that the Chicago City Railroad Company was permitted to use near the stockyards and on State Street from midnight to 5:00 a.m. to clear snow dumped during Chicago's brutal winters. The city council voted to allow their use on main lines during the Great Epizootic, but the number of dummies available was negligible and they provided little help in the absence of streetcars and omnibuses, though they were more helpful in the suburbs. Perhaps most important, the steam dummies demonstrated to politicians, city planners, and Chicagoans that machines don't get sick.[2]

For the year before the horse flu's arrival, Chicago had been abuzz with the visceral sensations of reconstructing a metropolis. Everywhere one turned, something was being razed, hauled away, or built. Then, when the Great Epizootic struck, virtually overnight, the city became eerily quiet and still. "Building operations have, of course, been almost entirely suspended," read a story in the *Tribune*, "as material cannot be supplied." Notices about construction interruptions became common, such as the one published in the *Evening Mail* about a delay in building a 1,500-seat public hall on Clark Street, informing readers that "work is to be commenced as soon as the 'epizootic' abates." Without horse power to lug construction materials into and around the city, the rebuilding of Chicago slowed to a crawl. As did commerce.

"The wide, open space around Market Street which is usually the waiting ground for hundreds of drays and carts and public wagons was pretty much deserted," described an *Inter Ocean* piece. But men with still-healthy horses lined their pockets. In a piece titled "The Hackman's Harvest," the *Evening Post* railed against draymen and hackmen who "during the interregnum in horse car and omnibus rule" appeared to believe "it is their privilege, if not their right, to require just what terms they please and to refuse, or condescend to serve just whom they please." Rather than be hustled by draymen lucky enough to own an uninfected horse, the owners of the Field & Leiter department store—whose building had been incinerated in the Great Chicago Fire and then rebuilt at Madison and Market Streets—turned to oxen to move goods to and from their store.

Like Field & Leiter, many stores, small and large, had trouble stocking their shelves and delivering their products. A few smaller companies even hitched dog teams to their wagons to move merchandise. Grocers, butchers, and milkmen, the *Evening Mail* told readers, "employ men, boys and themselves to deliver their orders." W. W. Chandler, an agent for the Star Union Freight Company, ran an ad in the *Evening Post* that pulled no punches: "GENTS: it is entirely unnecessary for me to explain why the Star Union Line has not delivered you any goods by our team since last Friday. Our usually crowded, and busy streets, now deserted, proclaim in language not to be misunderstood that the 'EPIZOOTIC' has arrived." The ad went on to proclaim that the Star Union Freight Company had goods for hundreds of businesses piled up and ready to deliver—but no horses to draw their wagons.

As the Great Epizootic continued, staples of all sorts were harder and harder to come by unless people lived in walking distance from a shop that just happened to be well stocked. Slaughterhouses near the stockyards had trouble securing horse power to bring meat into the city, with the weekly "provisions report" in the *Inter Ocean* noting that "this branch of trade has, in common with all others, suffered severely by the prevalence of the horse disease." On top of that, farmers from nearby often had barns full of grains, fruits, and vegetables, for lack of healthy horses to transport them into the city. Hungry and cold, particularly after all the residents of the city had gone through the prior year, was an especially painful combination, but in November, Chicago starts to get frigid, and the winter of

1872 saw November temperatures dip into the mid-20s (F°) just as the *Tribune* was warning that "the coal yards and their patrons are severely at their wits end in lack of hauling facilities." Those haulers were down with the horse flu.

Chicagoans, as had their fellow city dwellers in New York City and Boston, came to learn just how much they took for granted about life in a horse-powered city. It wasn't just food, transportation, leisure, and access to all kinds of goods that depended on equine power. Suddenly, mail service, when it was available, was snail-like. Almost all the fire horses were sick, which was particularly disconcerting. When the *Tribune* reported that the fire chief's horse died on November 8, it was hard for Chicagoans not to think back to what they were doing their best to move past. The court system, at the highest levels no less, also relied on horses, albeit in indirect ways. "Nine Bailiffs and three Judges are supposed to be doing duty in the Superior Court," a short article in the *Tribune* read, "but the fact is that 3 Bailiffs and 1 Judge are all that the epizootic [and] human illness . . . have left for the purpose." A young couple, on what should have been the happiest day of their lives, had no horses to take them to the train station so they could head out on their honeymoon. Fortunately, "their male friends harnessed themselves into the carriage from which the horses had dropped and drew the bridal pair to the depot amid much merriment."

Rich or poor, everyone was affected to one degree or another. "The horse-disease extends its influence to society and to the education of the young," read a *Tribune* story. "Its effects on the former was noticed at a fashionable westside reception on Saturday night when the drivers of the horses epizootically inclined distinctly informed the guests they must break up at 12:00 or go home on foot. The company was then immediately dispersed." Even the dead were inconvenienced, as both of the horses of a coroner named Stephens were down with the flu and he couldn't "procure a rig anywhere in the city for love or money [and] he is compelled to have a few of the corpses sat on by someone else."

Perhaps because of all they had been through since the Great Chicago Fire, Chicagoans mustered a richer black sense of humor about the Great Epizootic than did the residents of New York City or Boston. The *Tribune* noted that even in the face of all the chaos the epidemic was wreaking, at least "there has been a notable decrease in the number of depredations by horse thieves in the city since the advent of the epizootic." The *Evening Post* couldn't resist an epizootic swipe at anyone who challenged General Grant for the presidency, quipping that "the worst Hor'ce distemper—the summer madness that nominated Greeley."[3]

About three weeks after the Great Epizootic had begun, the *Evening Post* ran a story titled "Passing Away: The Disease Which has Prevailed Among the Horses is Gradually Disappearing." By that point more than one thousand horses—about 4 percent of horses in the city—had died as a result of the Great Epizootic, but at last, Chicagoans could breathe something of a sigh of relief as the worst seemed to have passed. While they were now commiserating with Bostonians over *their* Great Fire, the *Tribune* multipart headline of November 13, 1872, was more than comforting: "Convalescent: Our Equine Patients Recovering Rapidly: The Horse Disease Nearly Played Out in Chicago: Very General Resumption of Business: The Streetcars and Stages Running Again." Perhaps the surest sign the city was ready to move on was over at the newly renovated Myer's Opera House, where people could watch a full-blown comedic assault on the epidemic. For it was at Myer's Opera House, on "a new bill brimming over with rich . . . business," that Chicagoans could soak in a new burlesque show, "a roaring extravaganza," called *The Epizootic*, with the lead actor playing the role of Dr. Hippywithers. "Those who have mourned over the 'epizootic,'" the *Tribune* advised, "should go now and laugh at it."

It was time for Chicago to recommence rebuilding.[4]

# 9
# The Newsboys Sing "Old Epizook's Goin' Around and Around"

The Great Epizootic was causing havoc with the very fiber of America's urban centers. Philadelphia was no exception, and as with so many cities crippled by the epidemic, the effect on the pulse of the city—the streetcars that people relied on to get to work, family, friends, the theater, the library, and more—was profound. But streetcars had their own unique history in Philadelphia. When Ellen Anderson, the courageous African American woman we met in New York City, had stood up to the Eighth Avenue streetcar line for her right to board, it was an isolated incident; but in Philadelphia, streetcars had emerged as full-fledged stages for early civil rights protests. In the early to mid-1860s, not only were there many instances when brave African Americans stood up to the discriminatory, segregationist policies of the city streetcar lines, but also these actions generated a swell of support. Lucretia Mott spoke up, as did abolitionist William D. Kelley, a radical Republican, who published a well-circulated, scathing attack on streetcar segregation in a pamphlet entitled *Why Colored People in Philadelphia Are Excluded from the Street Cars*.

William Still, an African American whose parents were born into slavery but purchased their son's freedom, was the most prominent of Philadelphia's Free Black community to take a vocal stand against the racist policies of the dozens of streetcar lines in the city. Still is best known to history not just as an active agent on the underground railroad, but also as the author of a self-published book, *Underground Railroad,* one of the few first-person accounts of the dangerous passage from slavery to freedom by an African American. In 1859, Still and his colleagues stood in No. 25 Merchants' Exchange and presented to the board members of streetcar lines "A Petition for the Colored People of Philadelphia to Ride in the Cars," signed by 359 prominent African American and White citizens of Philadelphia. The petition began: "The colored citizens of Philadelphia suffer very serious inconvenience and hardship, daily, by being excluded from riding in the city passenger cars. It continued: "Colored people pay more taxes here than is paid by the same class in any other northern city. . . . Therefore, the undersigned respectfully petition that the various Boards of the city passenger cars rescind the rules indiscriminately excluding colored persons from the inside of the cars." This petition was then published under the title "Colored People and the Cars" in two Philadelphia newspapers, the *North American* and the *United States Gazette.*

There were also editorials aplenty on the streetcar policy as it related to African Americans. "The public highways are not the exclusive property of persons of any particular shade of complexion and the railway companies . . . have no more right to refuse to carry well-behaved colored passengers, than the owners of the Penrose Ferry Bridge have a right to forbid black men crossing the Schuylkill [River] by means of their structure," opined the *Philadelphia Press.* "The right of transit is not the exclusive privilege of any particular class, and the common carrier has no more right to refuse the privileges . . . to the colored passenger, than he has . . . [to] deprive him of the common blessings of light, air and water." At long last, in 1867, just five years before the Great Epizootic, the Pennsylvania state legislature banned streetcars from discriminating against African American riders.[1]

When the Great Epizootic was first recorded in Philadelphia on October 26, 1872, the population of the city hovered at 675,000, making it the second-largest metropolis in the country, behind New York City. Of the working population, some eight thousand people were either employees of the streetcar lines, working in stables and liveries as teamsters or drayman, or were somehow or another directly reliant on horse power for their living. Tens of thousands of others, working in the manufacturing and constructions sectors, also depended on horses, if only slightly less directly. As in other large cities, theater, music, and entertainment also depended on horse power. Philadelphia was home to Fairmount Park, rival to New York City's Central Park, which spanned almost three thousand acres and soon would host the 1876 Centennial Exhibit as well as house the Philadelphia Zoo: in 1872 alone, 26,500 Philadelphians brought their horses to the park for a gallop, trot, or leisurely stroll.

In early November, less than a week after it was detected, the Great Epizootic was raging, and fear was running high in Philadelphia. "It begins to be painfully apparent that the hopes, so earnestly entertained, that this city would escape the tribulation incident to the visit of the horse pestilence in other places were delusive," read a November 1 *Philadelphia Inquirer* story. "The dreaded disorder has in its march invaded the city and each day in the coming week will doubtless discover more and more victims of the malady until the comfort and convenience of all classes of people shall have most seriously interfered with all branches of business prostrated."

Despite their best efforts, it certainly seemed the malady in question had befallen the horses of the stable of the West Philadelphia Passenger Railway. "The establishment had been thoroughly renovated and disinfected in order to arrest the course of the disease," a *Philadelphia Inquirer* piece read, "but the precaution proved of little avail, as the disorder, stimulated by the cold penetrating atmosphere of the many days of the last week continue[s] to advance rapidly." Half of the 500 horses in the stable "were infected by slight symptoms of the plague." The *Inquirer* reported that the superintendent of another streetcar stable told a reporter "it was his opinion that half the car stables in the city were affected by the plague, but the companies did not desire the fact to be known and consequently were keeping the matter quiet . . . [though] he [himself] did not believe in smothering the facts."

The city council approved the use of steam dummies, but there wasn't time to have them brought in, and, in any case, the cobblestone streets and special rails in place would have made finding such dummies that would work exceptionally difficult. A few streetcars were being lugged by people, but the number was negligible. By November 1, the *Inquirer* reported that "nearly every railway company in the city" had significantly decreased the number of streetcars running their routes.

At the height of the epidemic, with the number of healthy horses running low, owners of streetcar lines decided that to generate revenue, they would pack between fifty and sixty people onto a streetcar meant to hold twenty and hope the healthy horses pulling those cars stayed healthy. When some protested, owners of the streetcar lines countered that the exercise of pulling overloaded cars was good for the horses and would help them fend off illness. The Pennsylvania Society for the Prevention of Cruelty to Animals (PSPCA), in turn, issued a warning to the streetcar companies. "I have directed the agents of the society to visit the depots with the passenger cars, to request them not to use any horse that shows symptoms of the prevailing epidemic," A. L. Elwyn, the president of the PSPCA wrote to streetcar company presidents. "I wish to avoid making arrests and therefore give this notice. After this notice any horse found at work will be removed from the cars and if seen again in use then the law must take its course."

Elwyn did not have the sway of Henry Bergh. The superintendent of the Race and Vine Street Streetcar Company told an *Inquirer* reporter "that so long as a horse will take his usual allowance of feed he will make that horse make trips." When an *Inquirer* reporter visited the Market Street line and the Green and Coates line, he was told something similar. And when horses at these stables, and elsewhere, came down with the flu, there was, as there so often was, talk of miasma: "On Tuesday night and Wednesday morning the disease attacked all the roads simultaneously," reported the *Inquirer*, "and from this it was evidently strong in the atmosphere at the time."

Draymen were even more brazen with the use of their sick horses than the streetcar companies. A *Philadelphia Inquirer* article with the headline "Our Patient Toilers," and the subheadline "Driving Horses to Death for Dollars" told of "hundreds of the poor [horses] that were hauling merchandise through the city yesterday . . . scarcely able to drag the empty vehicle to which they were harnessed and yet their owners allured, no doubt, by the inordinate prices paid for hauling, permitted the weak creatures to be laden heavily." In one case, a drayman had been paid an outrageous $8 to haul some goods away from Market Square. The *Inquirer* thought that the calculus of these mercenary draymen was off, and that, in time, greed might stop them from forcing sick horses into labor: "One thing will doubtless bring these unconscionable persons to a sense of the penny-wise-pound-foolish nature of their action: their horses will have either been killed outright or debilitated at five times the pace they would have been had a humane course been adopted and their infected horses and mares had been properly treated."

Even in the face of the rapacious draymen forcing sick horses to work, the Great Epizootic was slowing the movement of goods. "Coal, wood and flour dealers are getting less and less able to deliver orders to their customers," the *Inquirer* informed its readers. As for the delivery of milk and baked goods, the situation was equally dire, so much so that "truckers who depend upon their own muscles to deliver the respective articles of consumption to the consumer increased materially."[2]

As we've seen with Albany, Syracuse, and other cities (both in the United States and Canada), it wasn't only metropolises like New York City, Boston, Chicago, and Philadelphia that were victims. In the northeastern United States, few cities were spared. Burlington, Vermont, reported cases of horse flu on October 26. Bangor, Maine, was hit about the same time, as were fellow Maine cities Portland, Augusta, Bath, and Topsham. Both the city of Providence and little Block Island, Rhode Island, popular with vacationers, reported cases in the third week of October. The third and fourth weeks of October also saw the Great Epizootic arrive in Hartford, New Haven, Waterbury, and Rockville, Connecticut.

Further east still, the British were getting anxious. There was an active transatlantic trade between the United Kingdom and the United States—particularly via port cities like Liverpool and New York—and there was growing concern that it was only a matter of time before the Great Epizootic would make its way across the Atlantic. Between October 28 and November 21, newspapers in London, Liverpool, Leeds, Ipswich, New Castle, Bristol, Chester, Coventry, Sheffield, Derby, and Ormskirk ran stories about the Great Epizootic, particularly on the havoc it was creating in Toronto and New York City.

On November 13, the *Devon and Exeter Gazette* published a story that "three horses said to be affected with the epizootic have just been landed in Liverpool from New York and we are informed that two of the three are in good health, the other has only a slight cough." Soon confusion and contradictory opinions reigned supreme. The *Lancet*, Britain's leading medical journal, reported that the number of cases of horse flu in the country was generally low, except around Sheffield, and that the symptoms displayed by ailing horses were not the same as those displayed in America. The other major medical journal in the United Kingdom, the *British Medical Journal*, wrote the symptoms seen in Devonshire may be indicative of the spread of the disease across the Atlantic.

Unlike their English counterparts, the Irish took a proactive stand. Following suit with the orders that United States Secretary of the Treasury George Boutwell had issued banning the importation of horses from Canada "so long as the existing horse disease prevails [there]," at a meeting of the Privy Council in Dublin Castle on November 5, the Irish government issued an order prohibiting the landing at any port in Ireland of any "horse, mare, gelding, foal, filly, ass, mule or jennet from, or which, at any time after the 30th September, has been in any part of America."

In the end, Britain was spared the Great Epizootic, though the *Coventry Standard* may have been wise to warn readers that "though the cry may be 'Wolf!' at present, the example lately . . . in America shows that we should be prepared for such a catastrophe."[3]

Back in the United States, the Great Epizootic marched south and west. On its southern path it arrived in Charleston, South Carolina, which was at last recovering economically from the ravages of the Civil War, sometime in the last week of October 1872. By November 4, the *Charleston Daily News* was telling its readers that "the horse malady has appeared in the city of Charleston to an alarming extent, and that a majority of the work horses . . . of the city are more or less affected," and then proceeded to produce a long list of stables, including the stables of the city's streetcar lines, with horses suffering from the flu.

On November 6, one day after people had gone to the polls to cast their votes for president of the United States, the *Charleston Daily News* began their story on the epizootic: "The excitement in regard to the rapidly spreading horse disease almost overshadowed the interest felt in the general elections yesterday." That same story noted that almost all the horses at the Citadel Military College were down and also included a letter from a foreman at the City Railroad informing the people of Charleston that his company would immediately pull sick horses from their lines "in order that each and every case of sickness which may occur among the horses . . . [be] cared for in the shortest time possible." In an attempt to stave off what some in the press were calling "the mysterious Northern malady," lame duck Governor Robert Kingston Scott imposed a quarantine on all horses entering the state.

A port city, Charleston relied on horse power to move goods to and from its wharves. The *Daily News* reported that the first cases were "as is very natural, among animals employed by draymen hauling the heavy freights that have been passing through Charleston [ports] during the past few weeks." At the start of the outbreak, drayman, and those who employed them, were not about to allow their horses' coughing and runny noses to empty their pockets. "Many of the drivers continue to cruelly neglect the animals under their charge. . . . It was noticed that many horses and mules were allowed to stand on the wharves and other places where they were exposed to wind and drafts without any covering, and while sweating from hard working," the *Daily News* told its readers. "This would be shameful negligence at any time, but when the horse is suffering from a cough and a steady rain is falling amounts to criminal recklessness." Other horses owned by drayman and streetcar companies were just too sick to work, no matter how hard-pressed, and that "lack of horses," the *Daily News* continued, was "causing serious inconvenience to merchants, shippers & c." The next day saw "staid and venerable merchants who have not walked home to dinner for many a year . . . hurrying laboriously along with ill-disguised manifestations of disgust in the face and general bearing."

The Great Epizootic hit hard and fast, as was its wont. About three weeks after it first was noted in Charleston, most infected horses were almost recovered and business and transport were getting back to normal. Once the epidemic had passed through the city, epizootic poets in Charleston took pen to paper. In late November, one of these poets captured what the city had been through on the pages of the *Daily News*:

Not a sound was heard on the silent street
  As home from the concert we hurried
For we found not a street car, carriage or bus
  And we felt considerably worried
We hailed the driver we used to know
  And hurriedly asked the reason
He said as he sadly shook his head
  That the horses were all a 'sneezin
And so we silently went our way
  To the place where we kept our filly
We found her weeping at the eyes and nose
  And shaking as tho' were chilly
In vain we offered her tempting oats
  And a "bran mash" warm and smoking
She turned her head from the fragrant bait
  And swallowed as if she was choking
We shook down a bed of fresh new hay
  With a bag of corn for a bolster
And we gazed at the saddle overhead
  With the shining stirrups and holster
We thought as we looked in the glazing eye
  That her troubles would soon be over
Nevermore would she crop the graze
  Or graze on the Crimson Clover[4]

Perhaps because it was not the bustling port city that Charleston was, or perhaps because it was considerably smaller, Nashville, Tennessee, wasn't hit by the Great Epizootic until nearly two weeks after Charleston. On November 16, there was some uncertainty about whether the disease had reached the city, but no doubt that it would: "The epizootic, or horse disease, if not already among us," the *Tennessean* reported, "cannot be long diverted raging as it is." The paper added a word to the wise, imploring its readers to learn from other victims of the epidemic: "The experience in other localities, that the disease has been fearfully aggravated by overworking the horses, or, in fact working them at all, while suffering with distemper, should teach a humane lesson here."

By November 17, all the uncertainty was gone, as the *Nashville Union and American* made clear in its headline "THE EPIZOOTIC: The Horse Plague Reaches Nashville." The Spruce Street, North Nashville, and South Cherry Street streetcar lines were pulling cars from their routes for want of healthy horses, the paper reported. By November 19, all streetcar services in the city had shut down. As the epidemic raged, the mail was being delivered by wagons pulled by oxen and the fire horses were down with the flu. William Stockwell, chief of the Nashville Fire Department, called a meeting in City Hall and told everyone gathered that he had managed to replace all the sick horses with healthy ones from around the city. Still, he wasn't taking any chances. Chief Stockwell worried what might happen if the horses he secured, and indeed if all the horses in the city, became infected. He asked all those present to be part of a volunteer unit who, should a blaze break out, would pull the fire trucks in lieu of horses. Such a brigade was formed that very day.

Just as winter was approaching, the Great Epizootic was playing havoc with heating supplies in Nashville. "Owing to the presence of the disease here," the *Union and American* reported, "various coal companies have been unable to deliver coal . . . the cold weather increased the orders, and when it became known that delivery of fuel would not be continued, a rush for the coal yards was made and the scene was quite lively." In another *Union and American* story, the editors called out the blatant ruthlessness some were displaying toward sick horses. "Hacks are being run when the poor horses can hardly move along," the editors bemoaned, "and the cruelty evinced by the owners in thus persecuting dumb animals indicates how mighty is the love of filthy lucre and how far human nature may be prostituted for the acquirement of a penny."

On November 21, Nashville's mayor Thomas Kercheval had seen and heard enough and he took extraordinary action. In a public proclamation, published in the *Tennessean* and the *Republican*

*Banner*, the mayor instructed that "whereas, it is believed that the best and most prudent manner of preventing the spreading of the Epizootic in the city is to prevent the using of animals on our streets while infected with the disease . . . therefore all persons working, or in any manner using horses . . . which are affected with the Epizootic or 'horse disease' are hereby respectfully requested to cease the working or using of the same until they shall have recovered sufficiently from said disease. Persons found willingly persisting in using animals under the above circumstances will lay themselves liable to arrest."

As the epidemic progressed, it became harder and harder for farmers to find horses to bring their produce to markets in Nashville (and elsewhere in Tennessee) and food prices soared. "If the present rate of increase in prices is kept up for a week or two the scenes of the old Confederacy days will be revived," the *Union and American* warned it readers on November 24, "and we will see housekeepers on their way to the market carrying their money in a market basket and returning with purchases in their pocketbooks."

Fortunately, that worst-case scenario never came to be. By the end of November, the city's papers were calming the edgy nerves of Nashvillians. The streetcars were operating close to normal and "a larger number of horses appeared on the streets yesterday than has been seen since the advent of the hippophalgia," the *Tennessean* reported on November 28. "The condition of these horses is very encouraging when we consider that they have been afflicted with the disease for only two weeks, while most of the horses in northern cities could not be used in four to six weeks from the time of the inception of the disease. This we consider a good omen for all the remaining horses which have been prostrated by the eppihipic." Two days later, that same paper told its readers that "it is now generally conceded that the eppihipic has done its worst, that all horses attacked by it are rapidly recovering, and that no more fear need be entertained relative to its further spread."[5]

As the Great Epizootic moved west along its southerly route, it arrived in New Orleans during the last week of November. But the epidemic, and the questions that it raised about relying on horse power to run cities, was on the mind of the people of New Orleans even before horses began to fall ill. "People who had become unaccustomed to take any more account of the vehicular horse than of his vehicular driver as an important agent in our social economy, awoke with a painful shock to a sense of the inconveniences and losses which society would suffer if suddenly deprived of the services of the horse," began a November 9 opinion piece in New Orleans's *Times-Picayune*. "Upon reflection, it came to be asked whether too much was not staked upon the horse and depending upon him almost exclusively as the mode of power and the internal transportation of our great cities."

Eight days before the first cases of horse flu were recorded, New Orleans's mayor Benjamin Franklin Flanders reached out to the Board of Commerce, the Board of Health, the governor, and the city streetcar companies about the possibility of a preemptive quarantine of all horses entering the city. Though the mayor himself was keen on the idea, and was still awaiting feedback from the Board of Health, the *New Orleans Republican* reported that "the majority of the opinions [the mayor] received from the best informed sources were against the city authorities doing anything in the way of establishing a horse quarantine even if it were practical," and Flanders let the matter drop. Later, the Board of Health would castigate Flanders and his administration for this decision.

The *New Orleans Republican* and the *Times-Picayune* were the first to report cases of the Great Epizootic. On November 23, the *Republican* informed readers that "the epizootic or hippophalagia has reached our city, first by importation of affected horses brought here on the river steamboat, and we learned last evening that several unmistakable cases have developed here. Owners of horses will therefore take timely warning and look well to their animals." The *Times-Picayune* seemed markedly less concerned. Following a comment that the "Epizoot is here at last," they commenced with a series of puns, including "the community must bear its loss with equine-nimity"; "fishermen must not be afraid. It does not affect horse mackerel"; "it is said that the disease has not attacked horse chestnuts"; and "the man with the horse laugh is now studiously avoided." For their part, the Board of Health found nothing humorous at all in the situation and began inspecting stables across the city.

By November 27, the *Times-Picayune* was taking the epidemic much more seriously, telling readers that "today we have to record its general prevalence in the livery stables as well as among other horses . . . It is reported that all of the horses of the express company have been retired from service as being sick of the disease. . . . The line of [street]cars between Napolean Avenue and Carrollton has been suspended on account of the sickness of our horses used on that line." Soon the Barrone and Carrolton, Canal and Clairbore, Magazine Street and Prytania streetcar lines had been shut down entirely for lack of healthy horses. At fire houses across the city, every single horse had come down with the disease and three had died. The firefighters were stepping up, though, "with a degree of heroism worthy of all praise," the *Times-Picayune* reported. "Fifteen members nightly sleep in each of the engine houses to move their machines in case an alarm is sounded."

The Great Epizootic slowed commerce, but New Orleans had an advantage as a port city with a particular topography. "Our river front is so extended," a story in the *Times-Picayune* reassured readers, "that by free use of small barges and tugs, should all the draft animals in the city be affected, the sugar, molasses, cotton and western produce coming in can be readily moved and with very little additional cost." In fact, small barges and tugs helped, but only so much: "Cotton languishes on the Levee [and] sugar sweetens under the sheds," the *Times-Picayune* updated its readers a few days later. That same article noted that "even music refuses its humanizing influence because of the epizooty. Yesterday, we witnessed the novel sight of Lewis Grunewald's piano cart drawn through the streets by men with a Steinway grand piano in it."

As the epidemic in New Orleans began to cool down about two weeks after it began, the press began to take a philosophical perspective on the interrelations of man and horse. "Convalescent horses begin to appear on the street . . . they wear for the most part a self-satisfied look, as horses well aware of their importance and assured that the world had at last come to recognize their individual claims and the value of their race and to deal with them accordingly," the *Times-Picayune* opined. "There is also something of a pity in these languid glances they cast upon human pedestrians . . . who can never hope to be the objects of the tender consideration at the hands of their kind which is lavished upon the epizootized quadruped."

Children, for their part, did what children do: they turned what they were seeing adults worry about into a game. One creative little girl "cut up her mother's toilet cushion, ruined the carpet and nearly scalded a confiding young brother to death, treating him with a brand mash for a 'play epizootic.'"[6]

Even islands, like Galveston, Texas, four hundred miles west of New Orleans, were not spared the Great Epizootic. The people and press of Galveston had been keeping tabs on the epidemic as it spread south and west and some preemptive measures were put in place to try and spare the horses on the island. Appealing for supernatural intervention, the *Galveston Daily News* wrote on November 5: "Westward the star of the epizootic takes its way, and the best thing that we can do is to pray that it won't come to Texas." In contrast, Mayor Albert Somerville took concrete steps to stop the disease from entering via horses brought into the city's ports. On November 28, before any horses had fallen victim, the mayor wrote Morgan Steamer Line officials requesting that "you do not allow any horses or mules to be shipped by your line of steamers from New Orleans to this city." The officials agreed to comply.

Despite these efforts, in the last week of November 1872, the first cases of the disease were recorded in Galveston horses. By December 3, 90 percent of the horses and mules at the City Railway Company were sick and streetcar service had been suspended. The Galveston Superintendent of Streets reported that most of the horses that powered street cleaning carts were too sick to work and that he was trying "to negotiate for a yoke or two of oxen, to use in keeping the main streets clean during the epidemic." Others turned to oxen power as well, as merchandise, including several thousand bales of cotton, and countless barrels of fruits and nuts, lay on the docks with no horses to transport it into the city. As the horses of Galveston lay "all packed up and smothered under blankets and their mouths encased in bags of hot mash. . . . Their painful coughing and snortings unheeded save by their vigilant and careful nurses," Mayor Somervile again took action, waiving all fees normally assessed on businesses for using oxen in the city.

For the next few weeks, the ports were quieter, stores were not as well stocked as normal, and city life slowed, but not to the same extent as it had in New York City, Boston, Chicago, or Philadelphia, or even smaller cities like Charleston and Nashville.[7]

About a hundred miles east, but some nine hundred miles north of Galveston, the Great Epizootic was moving westward across the middle of the country. During the first few days of December it struck Lincoln, Nebraska, which had only become the state capital three years earlier. Small by Eastern city standards, Lincoln's two thousand five hundred or so inhabitants made it second—though a distant second—to Omaha in terms of population size in the state. Lincoln was keeping tabs on the disease and residents could only have felt an unease, at best, when the daily *Nebraska State Journal* picked up a story published in New York that began, "What will be the effect of even a temporary withdrawal of the horse power from the nation is a serious question to contemplate. Coal cannot be hauled from the mines to run locomotives, farmers cannot market their produce, boats cannot reach their destinations on the canals, bills drawn against western shipments will be protested, merchants will become alarmed at the nonappearance of customers, panic will seize the community."

The anxiety generated by such a picture was only magnified when the first report on the actual appearance of horse flu in the city was published in the *Nebraska State Journal* on December 2: "The first horse in the city . . . took the Epizootic last Thursday. . . . It is now estimated by a prominent horseman that there are no less than 125 cases in the city." For a city with 2,500 residents, that was no small number of sick horses. "The Newsboys Sing 'Old Epizook's Goin' Around and Around,'" was how the *Nebraska State Journal* described the city's plight. Still, Lincolnites might be warmed a bit when they read that despite sick horses, at least some staples would make it their way: "The horses of Wm. Parcell, the milkman, have got the epizootic. With the 'milk of human kindness' in his heart, he supplied his customers this morning from cans on a hand cart drawn by himself." But the milk of human kindness could only do so much, which became evident when a few days after learning of Mr. Parcell, residents of the city were told, "The 'Epizootic' has sent much consternation . . . and teams are scarce on Market Square." Furthermore, "the Epizootic is raging to such an extent in Lincoln that passengers are transported to and from the depot by ox power." And another bad omen: news from Omaha that "the [fire] engine horses . . . have the epizootic and orders have been given to every man to run to the engine house on the sounding of an alarm and pull the horse cart to the fire."[8]

The pace at which the Great Epizootic spread across the country slowed dramatically west of Lincoln, in part because of the relative dearth of steam railroad lines in the western half of the country and in part because the absolute number and density of horses was lower in the West, which meant fewer susceptible horses and fewer mass outbreaks. As a result, it was not until April that the Great Epizootic reached California, and later still Oregon and Washington (a territory at the time). On the numerous paths the Great Epizootic followed to California, the epidemic struck in Kansas, Colorado Territory, Dakota Territory, Wyoming Territory, New Mexico Territory, Arizona Territory, Utah Territory, and Nevada Territory. In time, the epidemic also reached Mexico and Guatemala, though relatively little is known about those outbreaks.

Streetcars were less common west of Nebraska, though they were found in cities like San Francisco and Salt Lake City, and so in that part of the country, the effect of the Great Epizootic on mass transportation was largely felt in the stagecoach lines. As always, the epidemic hurt commerce in a myriad of ways, many of which were different in the western part of the country because of the different industries in play. In Utah, for example, many gold and silver mines had to shut down temporarily for lack of horse power to move mined material both within the mine and to its next destination. Many of those mines lay in the remote Wasatch mountains, where miners depended on periodic shipments of food and other necessities. Deliveries of those materials were powered by horses and during the Great Epizootic, many of those horses were too sick to work. The mail too, often delivered by stage lines such as Wells Fargo, slowed to a dribble, and what did come was "conveyed by men on foot."

In San Francisco, "the hacking cough and wheeze which attends the epizootic was heard in all our neighborhoods yesterday," the *San Francisco Examiner* reported on April 23, "and the rumbling of wheels was far from being as loud as formerly. Indeed, it is thought that at least two thirds of our horses are now suffering from the epidemic." The following day, the *Examiner* told readers that the "fire department has many sick horses to care for, and there was no telling how much damage might be done were a large fire to break out in our present crippled condition . . . [and] the animals which belong to the coroner's office are all sick and unfit for use." With nary a horse healthy, "The livery-stables . . . refuse to let out any of their stock at any price," the *Daily Alta* reported. "Vegetable, milk, bread, and ice teams fail to make regular calls on their customers, and many are entirely neglected." As in other places, rogues were taking advantage of the situation: "Hackman . . . who are so fortunate as to have horses able to be up and about are making hay while the sun shines."

Even before the Great Epizootic arrived in California, a bizarre story was circulating about its origins, tied to the atrocious policies of the United States government toward Native Americans in the west. That story was being spread in the pages of the *Placer Herald* newspaper, published in what was known as the Mother Lode region—because of its gold deposits—in the Sierra Nevada mountains. On March 1, 1873, a month before the Great Epizootic reached the area, Joseph Walkup, editor of the *Placer Herald*, warned readers of what was coming their way. The cause of the epidemic that was heading to the mountains could be traced, Walkup claimed, to a massacre perpetrated by the United States government against the Native America Piikani Nation (sometimes called the Piegan Nation).

"The horse epizootic," Walkup told his readers "is traced to General Sheridan." The same General Sheridan who, among other things, was so integrally involved with the Great Chicago Fire of 1871. Walkup's argument centered on what is known as the Marias Massacre (sometimes called the Baker Massacre), which took place three years earlier, on January 23, 1870. As part of the Indian Wars, General Sheridan had ordered Major Eugene Baker to lead the 2nd U.S. Cavalry to attack a Piikani village that sat alongside the Marias River in what was then Montana Territory. The intended victims were members of Mountain Chief's band of Piikani, but instead, an inebriated Baker mistakenly attacked a village led by a Piikani leader named Heavy Runner. Most of the men were hunting at the time of the attack, and many of the women, children, and elderly village members were suffering from smallpox. Baker and his troops killed at least 173 members of the village and likely more (accounts of the exact number vary). He then stole hundreds of horses, making it less likely that survivors of the attack would make it through the winter.

It was those stolen horses, Walkup claimed, that were to blame for the Great Epizootic. The stolen horses, Walker wrote, "were afflicted with a terrible distemper . . . but finally partially recovered." When they did recover, Walkup continued, the horses "were driven into Canada and sold, and from them the epizootic broke out, crossed over into the United States, and has well nigh swept all over the country." As Walkup saw it, this was a cautionary tale of "chickens coming home to roost with a vengeance."

There's no evidence at all to suggest that the horses stolen by Baker's troops had any connection to the Great Epizootic, or even that they were infected—in fact, the epidemic did not hit Native American Nations in the West with the severity it hit white communities, in all likelihood because Native Americans tended to let their horses run free, rather than housing them in corrals and barns, reducing the spread of the disease. What there is evidence of is Walker's strong Democratic politics and his anger at what he took to be the corruption of President Grant's Republican administration. Indeed, some historians have argued that blaming Sheridan, an ardent Republican himself, for initiating a chain of events leading to the Great Epizootic, may have been Walkup's indirect attempt to avenge the wrongs of President Grant.

The Great Epizootic and the Indian Wars were linked in a second way. Beginning in the winter of 1872, and lasting through the spring of 1873, the federal government spent $420,000 fighting what became known as the Modoc War, fought in both southern Oregon and Northern California. Since the mid-1860s, about half of the Modoc Nation lived on a reservation in the Cascade Mountains, but many Modoc continued to live on their native lands along the Lost River in Oregon, just north of California. In late 1872, the United States Cavalry was sent in to force the Modoc along Lost River near Oregon's to leave for the reservation.

The bloodshed began when Kintpuash, a Modoc leader also known as Captain Jack, fought cavalry troops near the Lost River on November 29, 1872. After that battle, the Modocs retreated to the Lava Fields, about 350 miles northeast of San Francisco. These beds were notoriously inhospitable and remote and most of the money that the United States government spent during the Modoc Wars there for the next four months was used to transport men and supplies to the Lava Fields. That transport was powered by horses, and during the spring of 1873, as the Great Epizootic was sweeping through Northern California, there were almost daily complaints reported from the cavalry units that they lacked the horses to keep supply chains strong.[9]

By June 1873, when it had for all intents and purposes burned itself out, the Great Epizootic had wreaked havoc in almost every state and territory of the United States. Detailed reports on that havoc came next in reports from three of the New Yorkers whom we met earlier: James Law, Adoniram Brown Judson, and Alexandre Liautard.

# 10
# The Reports

An early stage for the debate about whether the miasma theory or germ theory best explained the Great Epizootic that was turning metropolises into giant horse hospitals was a Massachusetts Board of Agriculture meeting held on December 4, 1872, while the epidemic was still wreaking havoc across the United States. A handful of doctors in the Boston area took up the germ theory position. But most concurred with classic miasma theory for the epidemic befalling the city's horses, despite the fact that weather reports from Boston show that the humidity and dampness prior to the outbreak—key components of the miasma theory—were both relatively low. October 1872 was drier than the October prior and the only rain in the week leading up to the first cases of horse flu being logged had been a sprinkle on October 18.

One of the people who the Massachusetts Board of Agriculture met with was Dr. Daniel Slade, a professor of applied zoology at Harvard's Bussey Agricultural Institution, which despite having been endowed by Benjamin Bussey in 1835 had only opened its doors in 1871. Slade stood before the Board of Agriculture and began by presenting "practical observations upon the epidemic which has prevailed so extensively among horses." After some cursory opening remarks describing the disease, Slade took what at that time was a scientifically conservative approach, noting that some diseases are caused by miasma, others by germs, and that "under circumstances not yet recognized, certain diseases may prevail in an epidemic form and prove contagious at one time, which at another time do not possess this character or do not make it manifest in the usual manner." Still, he danced around stating under which of these categories the Great Epizootic fell; but Dr. Joseph Bates, the next physician to go before the board, was more than happy to share his opinion on that matter.

When Board of Agriculture members asked whether the ongoing horse flu was contagious, Bates took up the flag of miasma theory: "I do not say that it is contagious; I say the horses are exposed to currents of atmosphere, not exposed to the disease in other horses." In response, one of the board members, a Mr. Goodman, who made it clear that he was not a physician or veterinarian, challenged Bates by telling him that "from the experience of a great many who have owned horses, as well as from the observations I have made, I am of the opinion . . . that this is in some respects a contagious disease." To illustrate, Goodman noted that outside of Boston, "We found that all the horses that went to the villages for purposes of business were attacked: all the stage horses that went to a town where the disease existed came back diseased, [but] all the horses which were kept in the stable and did not come in contact at all with other horses have been exempt." What's more, Goodman continued, "in stables where there are a hundred horses, if one is attacked they will all have it; but in the country districts it is only the horse that comes in contact with other horses that [is] affected by it that is [itself] attacked."

Bates was not deterred by Goodman's argument. He replied with an anecdote he thought made his position both clear and defensible. Imagine, a man and his horse come back from a city where a lot of horses are down with flu, and that man proceeds to proclaim, "Why, I drove my horse that was perfectly well to the city, day before yesterday, and when I got him home, he was attacked." Bates's reply, he contends, would be simple: "Well he might have been [attacked], because he had been exposed to the very atmospheric influences which have produced the disease throughout the land wherever it has spread."

Things might have ended there for the Board of Agriculture had it not been for the next man they interviewed. It was one thing for a board member like Goodman to make the case that horse flu was contagious. It was another altogether for Dr. Joseph Sturtevant, an expert on cattle, whose brother, Edward Lewis Sturtevant, was a graduate of Harvard Medical School and a well-respected agronomist, to plead that case. "I believe this disease is contagious," Sturtevant told the board, "and the seeming fact which leads me to believe that it is [is] that it breaks out just in those places where you would expect it to break out if it were contagious. The livery stables, and those places which have daily commutations with . . . Boston are found to have been visited by the disease."

This argument caught the attention of Dr. Farwell Fay, a board member, who wanted a more explicit definition of "contagious." Sturtevant was happy to provide it: "I mean by contagious disease a disease which is carried by contact with some virus . . . to the animal which receives it." Here, Sturtevant, like others of his day was using the word "virus" in the generic sense of a living contagion and thereby taking up a germ theory position. In case any doubt remained, Sturtevant gave an example: "If this matter which passes from the nose of a diseased horse is conveyed, either by the clothes of an individual or by contact with the horse, or is deposited upon the hitching-post, and thus comes into contact with another horse, and that horse receives the disease, I should call that evidence of contagion." Whether board members found the argument for miasma theory or the argument for germ theory more compelling, they didn't say, but both were clearly on display at their meeting. A case can be made that after hearing the arguments for both theories the board took a reasonable "Let's wait and see until we have more data" approach by remaining silent on horse flu's causation. Fortunately, that data would soon become available.[1]

Not long after the Great Epizootic finally burned itself out in late fall 1873, three reports on the epidemic were published, covering veterinary, medical, and public health issues. Two of those reports—one by Alexandre Liautard, chief veterinary surgeon to the New York College of Veterinary Surgeons and consulting veterinary surgeon to the Board of Health of the City of New York; and one by Adoniram Brown Judson, M.D., assistant city sanitary inspector in New York City—appeared in the 1873 *Annual Report of the Board of Health of the Health Department of the City of New York*. The third and longest, most detailed report was penned by Cornell Professor James Law and was published in the *Report of the Commissioner of Agriculture for the Year 1872*, which was printed by the Government Publishing House in Washington.

Alexandre Liautard's "Report on the Epizootic as It Appeared in New York" was the shortest of the three reports, running just four pages. It was also the least informative, coming off at times as cursory. Recall from our discussion of the Great Epizootic in New York City, that in real time (on October 26, 1872) Liautard had told the *New York Tribune* that the disease didn't seem to be "contagious and infective," and that he regretted that U.S. Secretary of the Treasury George Boutwell banned the importation of horses from Canada." That ban, as far as Liautard was concerned, was a waste of time and perhaps a waste of money because as he saw it, if he was wrong and the disease was contagious, it was already too late to stop it and if it wasn't contagious it would impede commerce for no just cause. In "Report on the Epizootic," Liautard gave a perfunctory review of the symptoms and treatment for horse flu, adding little to what had been in the plethora of newspaper reports during the epidemic.

As to whether he now thought the disease was best explained by miasma or germ theory, Liautard's sole comment comes after noting that abdominal, thoracic, or "nervous" symptoms—as opposed to the much more common but not as dangerous symptoms such as runny noses, coughs, and fever—appeared in those horses "exposed to unfavorable atmospheric influences," which might be a nod to miasma theory, but just as easily could be interpreted as suggesting that no more than exposure to bad weather makes sick horses even sicker. Liautard was still unwilling to take a strong position one way or the other.

In Liautard's defense, it is worth noting that his report immediately follows Judson's much longer report in the 1873 *Annual Report of the Board of Health*, and so it is possible that editors had tasked Liautard with being brief and to the point. While there is no hard proof of that, Liautard was a man who was passionate about veterinary medicine and would serve as president of the United States Veterinary Medical Association a few short years later. In that role, he showed himself to be, if anything, a man averse to the cursory and perfunctory, a speaker and writer with a tendency to run on at length when talking about his beloved field. Indeed, in the very first issue of *The American Veterinary Review*, published in 1877, Liautard was given the honor of writing the maiden article for what was meant to be the flagship journal of American veterinary medicine. In his hyperdetailed article, Liautard apologizes for running on at times. This background makes it clear that if given the opportunity, Liautard would likely have written a much more detailed report on the Great Epizootic, and taken a position more firmly in the camp of miasma or germ theory, than he did in the *Annual Report of the Board of Health*.[2]

Adoniram Brown Judson's twenty-five-page report, "Story and Course of the Epizoötic Among Horses upon the North American Continent in 1872–73," begins with a detailed medical geography

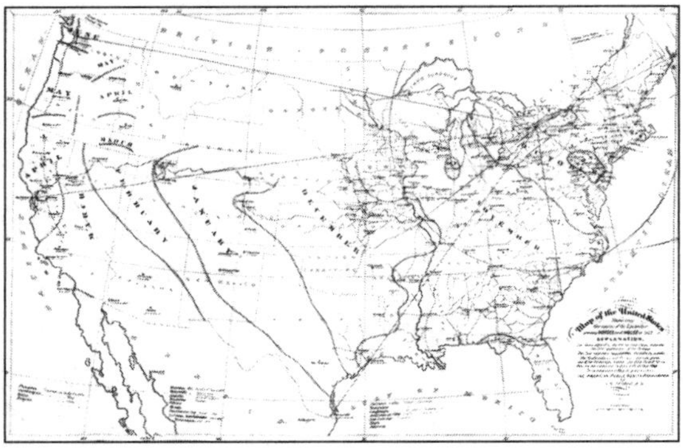

**Figure 10.1**

Judson Map of the Great Epizootic in the United States published in A. B. Judson, "Story and Course of the Epizoötic Among Horses upon the North American Continent in 1872–73," in *Public Health: Papers and Reports Presented at the Meeting for the American Public Health Association in the Year 1873* (New York: Hurd and Houghton, 1875).

of the Great Epizootic, with page after page of data on the first occurrence of horse flu in each state and territory in the United States as well as in Canada, and, to the extent possible, based on the *very* limited information available, in Cuba, Mexico, Guatemala, and El Salvador. In a separate though similar report that Judson penned on the Great Epizootic, based on a lecture he gave at the 1873 meeting of the American Public Health Association, he includes a six-page map that tracks the spread of the epidemic across the United States, listing approximate dates when horse flu reached different locations and overlaying major railway lines on the map when possible (figure 10.1).

From atop the mountain of post-hoc medical geography data he had amassed, Judson laid out three propositions centered on whether the miasma theory or the germ theory best explained the origin and spread of horse flu. "Proposition No. 1," he wrote in his *Annual Report of the Board of Health* article, was that "epizoötic influenza does not spread by virtue of any of the recognized atmospheric conditions of cold, heat, humidity, season, climate, or altitude." Here, Judson used the huge variability in temperature, humidity, and sea level seen across different locations where the Great Epizootic struck. "The disease prevailed and was propagated in the cold of a northern winter," as well as the warmer climate of the southern United States and Central America. It broke out "in the dry air of Minnesota, and in the moist air of the seaboard; at an altitude of five thousand feet above the sea . . . on the low levels of New Orleans, LA. (ten feet above sea-level), and Galveston, Texas (five feet above sea-level)."

Judson's second proposition played off the first: "Epizoötic influenza does not spread solely by virtue of unrecognized atmospheric conditions." This is a strange proposition for a doctor to make, as demonstrating that what is not known is not causing a phenomenon is untestable. Still, he tried. "During the prevalence of the disease, the opinion was expressed, by many thoughtful observers," Judson diplomatically begins his discussion of proposition 2, "that it was spreading through the air, or by virtue of some unknown atmospheric condition. In no other way did it seem possible to explain the sudden prostration of all or nearly all of the horses in a city or limited district."

Judson addresses this claim at two geographic scales. At the microgeographic scale, he took a conservative stand: "Investigation has not proved that the disease is not communicable through the air at short distances, and over limited areas," he writes of the possibility of miasma as a cause, though what investigations he is referring to are unclear. At the macrogeographic scale, Judson thought "the spread of the disease over the country is not solely, or chiefly, by virtue of unrecognized atmospheric conditions." His argument here is weak. "The irregularities in time and place," Judson proposes, "in the appearance of this disease, are so numerous and surprising that they cannot be classified or brought into harmony with any system of laws that bears any resemblance to the laws which govern the phenomena of any of the recognized atmospherical conditions." That part seems reasonable enough. Then, after listing a few of those "irregularities," including "that the disease appeared at places equally remote from its starting point at widely differing dates," he concludes that his second proposition is "logically proved, unless we are willing to believe in the existence of an unrecognized atmospheric condition whose phenomena are governed by laws which bear no relation of resemblance or analogy to the laws which govern the phenomena of the recognized atmospheric conditions." But nothing is logically proved here. Judson's logic is circular. He would have been better off moving straight from his first proposition to his third proposition, which was all about evidence *for* germ theory rather evidence *against* miasma theory.

Judson's third proposition was short and to the point: "Epizoötic influenza spreads by virtue of its communicability," meaning it was contagious, as germ theory proposes but miasma theory does not. Still, as Judson admits right from the start, "Experiments for the demonstration of this proposition have been wanting." Such experiments would be nice, he continues, but there were other methods to study the matter. "Admitting the present impossibility of demonstration," Judson writes, "we have abundant and convincing logical proof." That proof once again came from his overview of the medical geography of the Great Epizootic. On the one hand: "No place was exempt from the disease which was known to have been in communication, by means of horses or mules, with places in which the disease existed." On the other hand, if a location was "exempt from the disease [they] were so situated that the importation of horses or mules was in some of them impossible, and in others of them improbable," either because they were islands, such as Prince Edward Island in Canada and Key West, Florida, or because, as in the case, of Vancouver, they quarantined horses and mules. A final piece of evidence on this front came from the fact that "the disease passed rapidly over those regions in which the towns and cities are numerous and in frequent communication with each other," but with "comparative slowness over those regions in which the towns are less numerous and in less frequent communication with each other." None of Judson's data regarding his third proposition fits with miasma theory, but it fits germ theory like a glove.

Following Liautard and Judson's reports in the *Annual Report of the Board of Health* came other articles that record autopsies and blood and urine work from horses that had come down with the flu during the epidemic. The New York City Board of Health report's summary on the Great Epizootic closed with findings and recommendations. There was no question in board members' minds that the disease in question was horse flu and that "epizootic influenza is the counterpart of epidemic influenza," the latter referring to an epidemic of flu in humans. Furthermore, as far as horse flu was concerned, miasma theory could be rejected in favor of germ theory: "The result of the preceding investigations into the methods of propagation of Epizootic Influenza," the board stated in no uncertain terms, "is the presentation of abundant and convincing proof that it spreads from city to city, and from country to country, by virtue of its communicability, and regardless of atmospheric conditions."

The Board of Health knew all too well that "the effects of the prevalence of epizootic influenza on commerce, and on the general welfare of the community, [were] extremely injurious," and so they made recommendations based on lessons gleaned from the Great Epizootic. One recommendation to follow for future outbreaks seemed obvious now that germs, not miasma, were thought to be the cause of the epidemic: "It is evident that the introduction of epizootic influenza can be prevented by arresting communication, by means of horses and mules, with those places in which the disease exists." In addition to keeping healthy horses out of areas where animals were suffering from flu, the board also suggested when possible to establish a *cordon sanitaire*—cordoning off an area with infected animals and not permitting them to be moved elsewhere. Finally, the board recommended that all horses coming from overseas be quarantined at the slightest hint that influenza was prevalent from their country of origin.[3]

As far as presenting the case for germ theory, James Law did Judson one better in his own forty-five-page report "Influenza in Horses," which appeared in the *Report of the Commissioner of Agriculture for the Year 1872*. Law's report, like Judson's, included a tour de force of the medical geography of the Great Epizootic. But it also contained detailed information on the "atmospheric conditions" that miasmists turned to over and over again, as well as a synopsis of the results of a study Law had commissioned from a microscopist colleague. It was from that combination of medical geography, meteorological data, and microscopy from which Law drew many of his conclusions.

Law begins his report with a lengthy history of epidemics in humans and epizootics in animals from the time of the ancient Greeks onward, followed by a medical geography of the Great Epizootic that largely jibes with Judson's. Next, he spends page after page describing the symptoms of horse flu. In total these three sections make up a third—fifteen pages—of the report. But it's in the next section of the report, subtitled *Causes of the Influenza*, which itself runs twenty-four pages, where Law is in full form on the question of whether miasma or germ theory explains the Great Epizootic. As noted earlier in our discussions of the Great Epizootic in New York, on November 3, 1872, during the height of the epidemic in New York City, Law gave a lecture at Cooper Union in which he dismissed all the factors that miasma theory proponents thought were important in explaining the Great Epizootic. Still, much of that argument was based on principle: now, in *Report of the Commissioner of Agriculture for the Year 1872*, he had proof to back up all his assertions.

The *Causes of the Influenza* section opens with "the old doctrine of an epidemic constitution of the atmosphere has of late years been gradually waning," which may have been true in Law's circle of colleagues, but as we have seen over and over, is not reflective of the wide-scale acceptance of miasma theory by respected veterinarians and physicians in so many locations that were hit by the Great Epizootic. Law recognizes this reality to some extent, acknowledging that "more than any other epidemic malady, perhaps, has influenza retained its claim on an atmospheric causation." What's more, he makes clear, those claims assuredly were being applied to equine influenza, and in particular, that associated with the Great Epizootic. But "the epizootic of 1872 affords but the slenderest appearance of support to any of these [miasmic] hypotheses," Law begins his argument against such claims, continuing "neither soil nor elevation has materially affected it. The prevalence and mortality have been almost the same in the mountains of Vermont and New Hampshire as in the flat, malarious sea-coast of New Jersey, Maryland, and Virginia."

Law next turns to temperature, knowing that miasmists often argue that warmer temperatures create the "atmospheric disturbances" that cause disease. "The temperature has not exerted any marked influence," he argues, as "the disease has been general wherever it has reached." But Law digs deeper, providing pages of data and tables that demonstrate that not only was temperature unimportant as a cause, but also "sudden *changes* in weather or temperature," another mainstay of miasma theory, were not associated with the *origin* of the disease in the Toronto area in September 1872. One table that Law presents shows temperature, wind velocity, and barometric pressure, as well as the amount of rain and snow that fell every day of October 1872: indeed, the table shows each of those measurements gathered three times each day (6:00 a.m., 2:00 p.m., and 10:00 p.m.). As a comparison Law provides another table with corresponding data for every October from 1844 to 1872. "From these tables," Law argues, "it is manifest that there was no extraordinary state nor extreme changes in the weather during the whole of the month [of October 1872]."

Law was also very clear about what he was not saying. He understood that once a horse was infected, cold weather acting on an already compromised immune system could cause symptoms to be even worse, noting that "nasal and bronchial catarrhs often prevail extensively among horses, as among men, in connection with sudden and extreme variations of temperature, and especially in spring and autumn." But cold weather was not the *cause* of the infection. "Symptoms exist, . . . but contagium . . . of the malady is wanting," Law writes. It wasn't just that cold weather could make things worse but not be a cause: that was also true for many conditions that miasmists turned to as causes. "Overcrowding, with its . . . damp, vitiated air, has unquestionably been a main cause of the severity and complications of the disease in the large cities," Law notes, but then adds that there was nothing in the data to suggest "we should attach any importance to these in estimating the causes of this particular disorder."

Another miasmic bugaboo was fog, and Law tackles that in the same manner as he did other putative meteorological causes for the Great Epizootic: by employing data. Here, he grants that "remarkably acrid or fetid fogs have been observed to proceed or accompany some epidemics of influenza," such as in the 1727 and 1732 outbreaks of influenza in human populations in England, France, and Germany, but that was pure happenstance, he concludes, for "the present equine affection has neither been preceded nor attended by any such phenomenon. Fogs appeared on but three days, 6th, 11th, and 18th of September, 1872, whereas they existed on six days, 1st, 4th, 5th, 13th, 16th, and 19th of September, 1871." Again, he distinguishes between causes and aggravating factors. "Fogs and vapors, impregnated with sulphureous gases or other bad smelling or putrefying elements, would undoubtedly undermine the general health, and favor the diffusion of such a disease as influenza," writes Law, "but the origin and course of the present epizootic, like that of the majority on record, show clearly enough that no such condition is essential to development." In a similar light, there was no evidence that humidity was a cause of the disease. What then *was* the cause of the Great Epizootic? "Is there a specific contagion?" Law asks next. "This is manifestly a question of vital importance with reference to the influence of the above alleged causes."

For Law, the reason this was a question of vital importance was clear: if there was "a specific contagion, which may exist in the body of the sick animal, increase there, and be the means of communicating the malady to an indefinite number of sound stock, all our theorizing on noxious gases and putrid fogs, inclemencies and extreme vicissitudes of the weather, excess of ozone, magnetic disturbance, and the like, will be of small account." Law begins with a nifty bit of deductive reasoning. Usually outbreaks of equine flu are associated with outbreaks of human flu. That fits just as well with miasma theory—man and horse exposed to the same atmospheric conditions—as with germ theory: closely-related microbes infect man and horse. But in the Great Epizootic, no such outbreak occurred in humans. "If these resulted from atmospheric changes alone, how comes it that man has escaped now?" Law asks. "The explanation would be easy if the equine and the human malady were alike due to specific contagia, distinct from each other, but closely allied in their manifest results and in the conditions which favor their development or reproduction." Law's choice of words here is critical: *development* and *reproduction*. That is what germs do: organisms, not inorganic matter, develop and reproduce. "The only theory that will accord with the history of the malady and its steady increase and extension," Law concludes, "is that which recognizes the existence of a contagion, capable . . . of assimilating its appropriate food, of reproducing its elements, and of thereby increasing the area of the disease."

More evidence against miasma as a cause came from the abundant data Law presented on the spread of the disease, which shows how uninfected horses fall ill en masse once sick horses arrive in a city. "If it can be introduced in this way into a new locality, well out of the former area of the disease, and spread promptly from the imported sick animal as a center, it must be possessed of a specific contagium," Law argues. "Were the body merely charged with noxious gases, with decomposing organic matter . . . these agents would soon pass from the system and lose their noxious qualities by diffusion or decomposition." As a capper, he adds, "the presence of the sick animal would be no more injurious than a chemical laboratory, a putrid carcass, or an electric machine."

The only thing remaining was to try and figure out which contagion/microbe/germ caused the Great Epizootic. Law first addresses whether the horse flu might be caused by a fungus that spread in the nasal discharge of horses. Here, he turns to samples he collected at Cornell long before and then during the apex of the epizootic. Those samples were from both the nasal discharge of horses and "air obtained in stables." In samples collected at the height of the epidemic, he found a plethora of fungal spores both in the nasal discharge and in the air. But as Law is quick to point out, he found the same fungal spores, in the same abundance, long before the epizootic: they were always around and therefore not the cause of the disease. Law appended to his contribution reports from microscopists in Washington, D.C., and South Carolina, who attested to finding similar results, demonstrating that Law's findings were not unique to Ithaca, New York.

After ruling out fungi as the contagion, Law turns his attention to the "granules" that were found in sick horses. Here, he "recognizes in the granules, which exist abundantly in the diseased organs, the morbid agent capable of transmitting the disease." But how so? These granules were "microscopical particles of variable size and form," he argued that "often possess . . . a capacity to appropriate other material for the increase and multiplication of their substance. . . . [They] are reproduced with extraordinary rapidity in the substance of the diseased mucous membrane and at the expense of the vital elements, liquid and solid, of the body." Law comes tantalizing close to a modern viral explanation for causation. Still, those granules were visible under a microscope, but the virus that causes influenza was invisible in Law's day, as it is today (except using electron microscopy), and so why claim that he came very close to our understanding of viral causation? Following work by others, such as British microscopist Lionel Smith Beale, on the nature of granules, Law notes the granules found in horses sick with flu "either constituted the virulent principle or contained it." If the disease-causing agent—the "virulent principle"—was contained *within* granules, it might indeed be so small as to be undetectable. As far as Law was concerned, if that was the case, so be it, for all that mattered was "that there is a specific virulent element, which finds in the body of a susceptible animal the material essential to its growth, its unlimited reproduction, and extensive diffusion." And that much he was certain of. Law even uses the word "virus" in his report. While presenting the preceding argument, he notes again that while the "atmosphere" doesn't cause disease, disease-causing agents may be spread through the air: "The air may still be invoked as an important medium," Law notes, "through which the . . . virus or bioplasm . . . may be carried to long distances, and infect new animals and localities." Here, Law, like Sturtevant had done when he met with the Massachusetts Board of Agriculture, was using virus in the broad sense of a living contagion. In addition to laying out his arguments in the *Report of the Commissioner of Agriculture for the Year 1872*, Law published a somewhat condensed version of the same material in a venue that would have had a much broader readership in the medical community, *The Lens: A Quarterly Journal of Microscopy and the Allied Natural Sciences*.

Law's reports, as well as Judson's, provided strong evidence both for germ theory and against miasma theory. In particular, Law's point-by-point, empirically based data-driven analysis applied to the Great Epizootic was hard for miasmists to recover from. That isn't to say that miasma theory disappeared after the Great Epizootic. It didn't: certainly not in the newspapers where it appeared often for decades to come. But the frequency with which the medical community turned away from miasma and toward germ theory increased steadily thereafter. It is not possible to determine exactly when the coup de grâce for miasma theory occurred, but it is safe to say that miasma theory was all but rejected in medicine after Robert Koch, Friedrich Jakob Henle, Friederich Löffler, and Edwin Klebs laid out a series of postulates for how to definitively demonstrate the causal relationship between a disease and the microbe responsible for it.

Dubbed today "Koch's postulates"—though Koch himself never used the word "postulate" and the name does a disservice to Henle, Löffler, and Klebs—the version most often seen in the medical literature is based on a lecture Koch gave at the Tenth International Congress of Medicine in Berlin in 1890. Koch argued that to make the case that a given microbe causes a given disease, it must be shown that: 1. the pathogen must be present in all cases of the disease, 2. the pathogen must be isolated from an infected animal and grown in culture, and 3. the pathogen grown in culture must cause the disease in otherwise healthy organisms. Sometimes a fourth postulate—when the pathogen is reisolated from a formerly healthy but subsequently infected animal it must be shown to be identical to the original pathogen—is added to the list, though Koch himself does not discuss this.

No atmospheric disturbances, putrid emanations, noxious vapors, swamp gases, and the like: it was now taken as a given that the origin and spread of disease was explained by germ theory. Quarantine, *cordon sanitaire,* and many other medical measures that seemed beside the point if miasma was causing animal epidemics, but were obvious measures to take if germs were being transmitted from animal to animal, became standard practice. And in part, we have the Great Epizootic to thank for that.[4]

# 11
# The Broomstick Train

By the early 1870s, the residents, business leaders, and politicians of cities had come to take horse-powered urban transportation as a given. The Great Epizootic changed that. Everyone knew horses got sick, but to witness and live through an epidemic of horse flu that slowed life to a crawl for weeks and months was another matter altogether. And it could happen again. Mass transportation in cities would need to change. "There is a discomforting probability," the *Chicago Evening Post* wrote, "the horse plague in some form will trouble us more or less for years to come. This probability and the manifold inconveniences already experienced, have given a fresh start, in almost all our large cities, to discussion of means of more certain and rapid transportation."

In New York, as the Great Epizootic shut down the city, the *New York Herald,* the *New York Tribune,* and the *Nation* took the horse flu as an omen for worse things that might yet happen should another epizootic hit the horses of the city. "We are now for the first time forcibly reminded that a plague might break out among horses, as plagues have broken out among men," the *Nation* bemoaned, "which would sweep them away . . . and which would momentarily baffle science." The *Tribune* concurred, warning its readers that "it is quite possible that we may have in the future to confront an epidemic quite as unexpected, as extensive and a great deal more fatal." New Yorkers opening just about any newspaper in late October 1872 were seeing something similar, a call to begin thinking past the horse as the solution to all things urban: the *Herald* implored its readers and the powers that be in New York City to look forward, past the epidemic, and that there was "necessity of some solid and sufficient plan of rapid transport for the growing metropolis. Year after year we see buildings springing up in the upper part of the island and with the annual increase in the spreading out of the population comes new demands for some means of quick transportation." Similar calls were coming out of Boston, Chicago, and other large cities.

The Great Epizootic, the newspapers were suddenly claiming, screamed out that the shackles of the past needed to be broken. In the midst of the epidemic, a *Herald* story on November 3, 1872, proclaimed that we owed that to ourselves and our horses: "Humanity is slowly freeing itself from the thrall of the antique world," readers were told. "Steam has superseded water power in our mills and factories and taken the place of horse power in the wheat fields as well as the country roads. The present time seems advantageous to advocate a yet greater extension of its use, and if the present disorder shall aid in effecting this change it will certainly be ample return for all the suffering it has caused a lower race of animals."

In 1881, nearly a decade after the Great Epizootic "disorganized everything," the *New York Times* published an article titled "The Horse in Cities." The story opened with a quip—"some wise cynic has remarked that in this country cities are made by taking a horse out upon the plain and building around him"—and then got more serious, noting that it was still the case that city residents "depend[ed] . . . upon the animal for pleasure, traffic and locomotion." Part of what was holding up progress, beyond the horse as all things urban, the *Times* proposed, was that there were some people who confused means and ends. "The animal is only a means, notwithstanding that 'horse' men . . . and even the public generally get quite in the way of viewing him as the end. There is a groundless disposition to regard all roads as made for him [the horse] and to judge every proposed change as it proposes to affect him." Instead, the *Times* argued, "Let him [the horse] be considered indispensable, as an instrument, until some better instrument enables us to dispense with him."[1]

There had to be a better way, a better instrument of urban mass transport. One possible solution was to look up: to think about a mass transport system that ran not on the streets, but above them. In large metropolises like New York City, Boston, Chicago, and Philadelphia, daily life on the streets was congested enough as it was, and the notion of an elevated system of mass transportation—urban transit on stilts—opened a world of possibilities. The citizens and politicians of New York City began to understand that potential even before the Great Epizootic. In 1848, John Stevens published a proposal to construct an elevated rail system inside cast iron and glass, which would run along cables that drew their power from a generator at street level, and cruise along three miles over Broadway. In 1853, engineer James Swett also proposed an elevated railway, this one with a coal-burning steam locomotive running on elevated rails, and a passenger car that would hang below the rails. Though this never got past the drawing-board stage, Swett's idea as well as Stevens's and other early proposed elevated rails collectively got people looking up for solutions (figure 11.1).

In 1866, the New York state senate passed a resolution to create "a select committee . . . [including] the mayor of the City of New York, the state engineer . . . [to] report to the Senate the most advantageous and proper route or routes for a railway or railways suited to the rapid transportation with passengers from the upper to the lower portion of New York City." That committee met and determined that "public notice [be] given by means of advertisement

**Figure 11.1**
James Swett's 1853 proposed elevated railway for Broadway in New York City. From New York Public Library Digital Collections, accessed January 31, 2025, https://digitalcollections.nypl.org/items/510d47e0-cd32-a3d9-e040-e00a18064a99.

that parties desirous of submitting plans would have the opportunity to do so." Dozens of proposals were submitted.

The first elevated line to operate was the brainchild of Charles Harvey, a civil engineer, who had played a key role in the construction of the St. Mary's Falls Ship Canal connecting Lake Superior and Lake Huron. In 1866, when pitching the elevated rail idea to the powers that be in New York City, Harvey impressed them by offering to self-finance the construction of a quarter-mile test track in Lower Manhattan on Greenwich Street between the Battery and Morris Street. His railway car would move along the rail on a cable—designed by John Roebling, who would one day design the Brooklyn Bridge—and be driven by steam produced in steam engines placed underground every five hundred yards. If his idea failed, Harvey said, he'd pay to demolish the track; but if it didn't, he asked that the city consider funding his company—the West Side and Yonkers Patent Railway Company—to construct a full-blown elevated line to move New Yorkers around their city.

On December 7, 1867, Harvey successfully drove a small car on his test track and investors soon flocked his way. Six months later, the track had been extended another quarter of a mile to Cortland Street, and the mayor and governor took guest rides, chugging along at about ten miles per hour, and then glowing about the project. But they did not glow for long, as delays began to pile up. The *Times* wrote on November 6, 1869, that "this enterprise, which has now been before the public for about three years, and in the prosecution of which nearly half a million dollars is reported to have been expended, appears to be no nearer completion than it was a year ago" (figure 11.2).

But at long last, the elevated tracks were all laid along Greenwich Street and Ninth Avenue all the way up to Twenty-Ninth Street, and on February 14, 1870, the Ninth Avenue elevated line operating three passenger cars, each seating thirty-five passengers, opened to the public.

Only one car ran per hour, however. The public was underwhelmed. "This transit experiment is an obstruction which . . . men would not tolerate for an instant on Broadway," wrote the *Brooklyn Eagle*. In August 1870, the *Times* reported on the demise and death of the Ninth Avenue elevated line: the cable-driven railway, the paper noted, "ended its precarious existence. The amusement of running empty cars up and down the road, while it gratified the pride of the investors . . . was apparently too expensive."

The Ninth Avenue elevated line was no more, but the need for mass rapid transport had not changed: "There is in New York no more absorbing subject than the question of how to get up and down town safely and rapidly," read a piece in *Appletons' Journal of Literature, Science and Art*, "as almost the entire business and laboring portion of the population has to be transferred in the morning from the upper part of the island, or the suburbs beyond, to the lower part, and in the evening must be sent back again to their domiciles, the present means of travel are not only inadequate in extent, but are far too slow and cumbersome."

**Figure 11.2**
Cable car no. 1 at the Twenty-Ninth Street station of the West Side and Yonkers Patent Railway Company, 1869. From the Mid-Continent Railway Museum, https://www.midcontinent.org/rollingstock/CandS/dsp-passenger/nyelrail3.htm.

With rails laid, investors of the West Side and Yonkers Patent Railway Company were not ready to give up hope quite yet. They ousted Harvey as head of the company and reorganized as the Westside Patented Elevated Railway Company, which soon became the New York Elevated Railroad Company. Next, they received permission to run coal-powered steam cars, rather than cable-driven cars, along the already built rails. They also increased the number of cars and ran routes multiple times per hour. This time, the Ninth Avenue steam-powered elevated line was a success after commencing operation in April 1871. Soon the track was extended up to Forty-Second Street and then further north. The success of the Ninth Avenue elevated line, combined with the push for new means of transport after the Great Epizootic struck eighteen months later, spurred politicians and investors to action. The Sixth Avenue and Third Avenue elevated lines followed in 1878. Between September 1877 and September 1878, about fourteen million New Yorkers rode the elevated lines. The following six months matched that number. Soon the Second Avenue elevated line, which opened in 1880, increased ridership.

As New York City residents turned to the elevated rails for rapid transit, they turned away from horse-powered streetcars. Chicago followed suit in 1888, when the Chicago & South Side Rapid Transit Company was incorporated: the first elevated lines went into operation in 1892. On July 2, 1894, the Massachusetts Legislature incorporated the Boston Elevated Railway Company, to construct lines from historic downtown Boston to the suburbs. By the early 1900s, elevated trains were moving from coal-burning steam to electricity as their energy source.[2]

As big cities in America were building up to create an urban mass transport system whose engines weren't susceptible to disease, they also began looking down. They had done a bit of that before the Great Epizootic as well, though not quite as soon as London had. There, the Metropolitan Railway opened the first underground rail system, powered by steam locomotives burning coke (and soon thereafter coal), between Bishop's Road and Farringdon Street in London on January 10, 1863. The idea of an underground rail was seriously floated in New York City a year later, in 1864. That year, a bill was introduced in the New York state legislature to allow construction of an underground subway to run between the Battery and Central Park. That bill, championed by Hugh B. Willson, was defeated in the state senate. The *New York Times* excoriated the legislature for being a pawn of vested interests. "The defeat of the Underground Railroad in the Legislature . . . is a fresh and alarming illustration of the enormous power over our property and comfort which is now wielded by the omnibus proprietors, [horse] railroad corporations and political jobbers," the paper stated on April 11. "The Legislature on which we have been in the habit of relying seems to have surrendered itself completely to their hands. The Underground Railroad promised to be an immense boon to the city. It was perfectly feasible; a similar road has been in successful operation in London for one year."

The defeat of Willson's bill didn't stop Alfred Ely Beach from looking down for urban mass transport solutions years before the

Great Epizootic. Beach had worked at the *New York Sun*, after his father purchased that newspaper, and in 1846 he bought the fledging magazine *Scientific American* from Rufus Porter and took on the role of editor. Beach was an inveterate inventor and tinkerer: the year after he purchased *Scientific American*, he applied for his first patent—for a new typewriter—and at New York City's Crystal Palace exhibition in 1853, he touted a new machine that would produce raised type to help the blind read. Beach had also patented a number of devices using pneumatic tubes for mail transport. In the late 1860s, he proposed a pneumatically-powered underground mass transport system for New York City in which cars were moved along tunnels by air pressure.

In 1867, Beach built a small-scale model of a pneumatic subway. A thirty-foot-long wood tube, in which a single car was moved by the air pressure created by a giant fan, was up and running at the October 1867 exhibit of the American Institution, held at the massive Fourteenth Street Armory. More than one hundred thousand people swarmed in for a chance to take a test ride. New Yorkers were clearly enthralled, but Boss Tweed and his Tammany gang were getting kickbacks from horse-powered streetcars and omnibuses and were tied to the potential profits from planned elevated rails, and refused to let Beach take the next step and build a pilot underground pneumatic railroad.

Undeterred, under the guise of experimenting with pneumatic tubes to power the mail, Beach secretly built a 100-yard-long pneumatic railroad, powered by a 100-horsepower fan, under Broadway, running between Warren and Murray Streets. This was no easy matter. To keep the project under wraps, Beach couldn't dig long trenches on the streets, but instead constructed a tunneling shield, powered by hydraulic rams, predating similar techniques that would become more common over the next few decades. When Beach unveiled his new system to the public on February 26, 1870, they again loved it: during its first year of operation, 400,000 people paid the fare of twenty-five cents for the experience of sitting in a car blown back and forth between the stations at Warren and Murray Streets. Tweed's ring was even angrier than before, which combined with Beach's endless troubles garnering funds—made all the more difficult by the financial panic of 1873—doomed a full-scale underground pneumatic railroad project (figure 11.3).

Pneumatic-powered underground subways made little headway, but the need for mass urban transport, which had kicked into high gear after the Great Epizootic demonstrated just how vulnerable horse-powered mass transport systems were. Demand for alternatives had only grown greater and there were other new ideas for underground solutions to move large numbers of people around expanding cities.

Boston and New York raced to see who could dig down and create a full-blown underground mass transit system first. Boston won, hands down, opening their subway line more than seven years ahead of New York City. At 6:00 a.m. on September 1, 1897, 100 people boarded the Tremont Subway line running between the Park Street and Boylston Stations. That day would see more than one hundred thousand Bostonians and visitors take the three-minute trip between the two stations. The era of the American electric subway had begun. Seven years later, the second chapter unfolded in New York City. At 7:00 p.m. on October 27, 1904, after a day of various ceremonies, subway service in New York City began with the inaugural run going from City Hall up to 145th Street and Broadway. The fare was a nickel, and the 111,881 fares sold that day totaled $5,594.05, all of which the city donated to hospitals.

The *Times* considered this and other developments to be the death knell for horse-powered cities: "Though the horse is still the noble animal he was in our youth, when we wrote compositions about him at school . . . as a beast of burden and a motive of power, he has had his day," the paper editorialized. "Some persons will feel badly about this, but the progress of civilization must not be

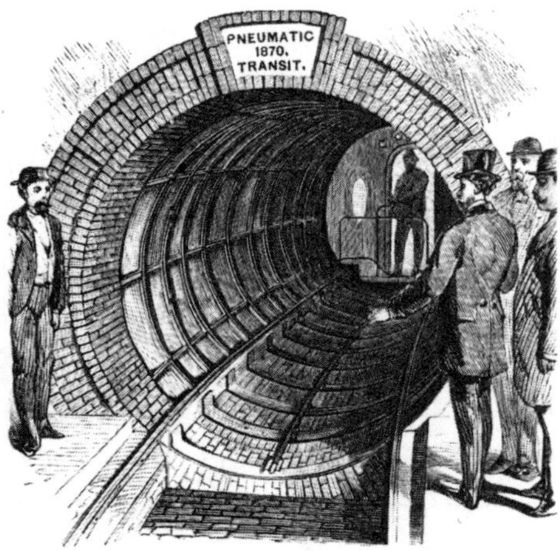


**Figure 11.3**

Beach's pneumatic railway. Top: From *Scientific American* 22, no. 10 (March 1870). Bottom: From *Illustrated Description of the Broadway Underground Railway, with Explanatory Details and Engravings* (New York: Beach Pneumatic Transit Co., 1872), Rare Book Division, New York Public Library Digital Collections, accessed October 29, 2025, https://digitalcollections.nypl.org/items/510d47dc-9aab-a3d9-e040-e00a18064a99.

checked for sentimental reasons . . . we can waste no sympathy on him. He has had his day and it has been a good long day too." The new vision of the future was the White City of the 1893 Columbian Exposition in Chicago, described by one historian as "monumental buildings, electric vehicles, and broad clean streets devoid of horse manure, powered by invisible, silent, effortless, clean electricity . . . a technological environment that was rationalized, efficient, sanitary, scientifically planned and managed, and devoid of the messiness of organic life."[3]

With elevated rails and subways, horse-powered transit was giving way to urban mass transit from above and below. New means of urban mass transport were, of course, also being tested at street level. As we've learned, during the Great Epizootic, New York City, Boston, Chicago, and Philadelphia turned to steam dummies as an alternative to horse-powered streetcars, but with very limited success. During the epidemic, there were other experiments with alternative urban mass transportation at the street level. As horses were falling ill in New Orleans, the *Times-Picayune* reported that their city was ahead of the curve on this front and its citizens should be proud that one of their own, Emile Lamm, had been working on a means of horseless mass urban transportation. Born in France, Lamm and his family had come to New Orleans when Emile was fourteen. Now an inventor as well as a dentist, Lamm had come up with what was called a "fireless engine," which in principle could navigate in cities without the use of steam or the need for rails embedded in the streets. The fireless engine used compressed air or liquid ammonia in lieu of steam power and Lamb had completed some successful pilot runs in a section of New Orleans known as Carrolton. "Perhaps we shall have it to boast, the *Times-Picayune* ended their story, "that the inventive genius of New Orleans has solved the grave problem involved in the unavailability, sooner or later, of equine muscle as a cheap and sufficient mode of power." And, indeed, Lamm's fireless engines did play a role in New Orleans transit in the later 1870s and then the 1880s. But during the Great Epizootic, it appears that just a single such fireless engine was available and put to use.

One breakthrough in street-level urban transit came soon after the Great Epizootic had ground horse power to a halt across America, when in August 1873 Andrew Smith Halladie introduced cable cars to navigate the notoriously hilly terrain of San Francisco. These cars "gripped" a cable slotted between rails, were powered by steam engines that moved the cable, and could travel about ten miles per hour. Others had developed the technology of the cable car, but Halladie is rightly credited with introducing the first practical cable car system to a major city in the United States.

Many cities adopted cable cars, either to supplement or as their primary mode of urban mass transport, with Chicago in the lead. On January 28, 1882, more than fifty thousand Chicagoans braved the city's notorious winter to line State Street, from Madison Street to Twenty-First Street, where a massive cable-car power station had been built. At 2:30 p.m. the city's first cable cars, carrying 500 passengers, traversed that path, as the Lyon and Healy Band stood on board regaling the passengers with celebratory tunes. Deemed "an unqualified success," by the *Tribune*, "the cars moved along beautifully and there was no jarring, such as experienced on the horsecars. Stops were easily and quickly made . . . horses on the street did not appear to notice the mysterious moving cars and the idea that this new mode of travel would frighten the animals was quickly dispelled." For the near future Chicago streets were ruled by cable cars, which at their peak ran on nearly fifty miles of tracks around the city.

The ultimate decline of cable cars as urban mass transport in Chicago and elsewhere came as a result of electrified urban mass transit, first on the streets, then under and above them. For all practical purposes, that began with Frank Sprague's electric trolley system. Even in his younger years, when he served in the U.S. Navy, Sprague was taken by electrical power. "Electricity had taken such a hold of my fancy," Sprague wrote later in life, ". . . that all during the cruise my messmates were made unwilling victims to listen to explanations of various inventions and to subscribe as witnesses to their understanding of them." After visiting the Paris Electrical Exhibition in 1881, Sprague began work as a technical assistant to Thomas Edison. In Menlo Park and elsewhere, Sprague not only worked on electrical lighting, but also helped develop an electric motor and convinced Edison to rely more on mathematical equations than trial and error in his lab.

In 1884, Sprague parted ways with Edison for New York City to open his own company, the Sprague Electric Railway & Motor Company. One of his many projects was developing an electric streetcar. After overcoming a series of engineering and design changes, including improving upon Belgium inventor Joseph Van Depoele's spring-loaded trolley pole through which electricity could flow from wires overhead to power a trolley, Sprague chose Richmond, Virginia, as his test grounds, because in that city "the road presented conditions of length, curves, character of roadbed and number of cars to be operated, which, if successfully overcome, would mark a new era in electric traction, and hence our ambition." After pilot runs, Sprague's company began laying twelve miles of trolley track in Richmond, and the trolley cars were up and running on February 8, 1888. Boston followed suit with its own electric trolley car system the next year. Eighty-year-old Oliver Wendall Holmes Sr. was so taken with the trolley car, particularly the trolley pole, that in 1890 he published "The Broomstick Train" in the *Atlantic Monthly*.

They came, of course, at their master's call,
The witches, the broomsticks, the cats, and all;
He led the hags to a railway train
The horses were trying to drag in vain.
"Now, then," says he, "you've had your fun,
And here are the cars you've got to run.
The driver may just unhitch his team,

We don't want horses, we don't want steam;
You may keep your old black cats to hug,
But the loaded train you've got to lug.
. . .
Often you've looked on a rushing train,
But just what moved it was not so plain.
It couldn't be those wires above,
For they could neither pull nor shove;
Where was the motor that made it go
You couldn't guess, but now you know.
Remember my rhymes when you ride again
On the rattling rail by the broomstick train.

Holmes wasn't the only one bewitched. The ascendency of the electric trolley car—soon referred to simply as the electric streetcar—was nothing short of breathtaking. In the five years following Sprague's trolleys taking to the streets of Richmond, almost every major city began the transition from horse-powered to cable-powered street-level mass transport. By 1902, just fourteen years after trolley cars began rolling on the streets of Richmond, about 95 percent of the close to twenty-thousand miles of streetcar tracks in the United States were electrified.

As the transition to electric-powered mass urban transport was occurring, there were any number of crossovers between trolley cars, elevated railways, and subways. Indeed, the very first *subway* car, rolling on Boston's Tremont Subway, was a trolley car. When trolley car no. 1752—displaying a yellow sign that read "Subway to Park Street," and stuffed to the brim with passengers who had queued up at 4:00 a.m. to board—began its route, it was *on* the streets before it descended down a ramp into the Tremont tunnel at Boylston Street, which explains why on the very day subway opened, the *Globe* ran a front-page story, titled "First Car Off the Earth." In the tunnel, overhead electric power lines were in place to keep the trolley car, now acting as a subway car, moving until it

**Figure 11.4**
A Mack truck used for sightseeing in Brooklyn, 1900. From the Mack Trucks Historical Museum.

reached the Park Street Station three minutes later. Soon electric-powered subways on some lines would operate below ground but, at times, emerge to soar above on elevated rails.

Gas-powered buses, first introduced in 1895 by Carl Benz, when his *Landauer* model traversed the route between Siegen to both Netphen and Deuz on March 18 of that year, provided yet another horseless form of mass transport (figure 11.4). In the United States, the earliest such buses, produced by the Mack Brothers Company in 1900, which could seat twenty and travel at a respectable twenty miles per hour, powered by a ten-horsepower engine, were used for sightseeing tours of Brooklyn. Starting around 1910, the rise of gasoline-powered trucks led the construction industry to turn away from the horse as a key cog in the process of powering the growth of cities. Of course, there was the rise of the automobile as well, though its effects, great as they were and would become, were not so much about *mass* transport or commerce early on (aside from commerce in the automobile industry itself).[4]

The most rapid and obvious decline of the horse in urban centers after the Great Epizootic was in their use as engines of mass transport. In other arenas, such as short-haul deliveries, particularly around docks, warehouse districts, and freight services it took longer before the horses disappeared as living, breathing engines of the city. And, of course, the rate at which both these declines in horse-powered transport occurred depended on the exact region of the country, with the change being quickest in the Northeast and cities like Chicago.

Gradually, over the decades, the day of horses powering cities came to an end. Jolted into high gear by fears generated by the Great Epizootic, technology, combined with the economics of alternative, faster, more reliable and more efficient mass transport that never fell prey to disease, had changed the face of city streets and the pace of life across the country. It also changed city dwellers' relationship with horses.

With no horses powering the cities, gone were the days when they were the heartbeat of a metropolis, even if they created an infernal stench by coating the streets with their manure and urine. Gone too, arguably, was the last link between city life and the natural world. True, there were parks and gardens offering urban flora and fauna. And dog ownership was on the rise. But the last visceral connection city dwellers had on a daily basis with another species was no more.

# Epilogue: From the Great Epizootic to the 1918 Flu Pandemic?

> I had a little bird, and its name was Enza.
> I opened the window and in-flu-enza.
>
> —A children's jump rope song from 1918

In the middle of November 1872, just as New York City was returning to normal, there was one last scare that the Great Epizootic shot through the heart of residents of the city: that their Thanksgiving meals were in jeopardy. In a November 18 article—"Is it the Horse Disease among the Poultry?"—the *New York Herald* told New Yorkers, who were more than ready for a festive gathering after the Great Epizootic, that "we have bad news for Thanksgiving [from] up the Hudson and from the Highlands and lowlands of this river back to the Susquehanna. . . . A terrible disease has broken out among poultry. . . . Chickens are dying from it by the hundreds and ducks geese and turkeys in the market are said to betray evidence of the same malady." If that wasn't scary enough, the story informed readers that the disease in question "appears to resemble the [horse] epizootic . . . should this pestilence continue to spread as rapidly as it has been developed, we can only hope that at least our beef, pork and mutton will be spared for Thanksgiving Day." The next day, the *Herald* suggested that perhaps their original report might have been too gloomy, but other papers were now bemoaning that in New York and Pennsylvania turkeys were "dying by the thousands!—Poor prospect for Thanksgiving Day."

The malady was dubbed "henfluenza" by most and Pferdhuhnkrankheit (horse-chicken-disease) by some in the German community. Unlike the rumors about transmission from horses to humans, there is little question that fowl were indeed coming down with flu-like symptoms and that the geography of the henfluenza outbreak mapped onto that of the Great Epizootic. It's hard to estimate how many turkeys, chickens, and geese died from henfluenza, but it was certainly hundreds, and perhaps many thousands: even if the latter, in the end, it had little effect, for as the holiday approached "the hooks of the stalls fairly strained with the loads of luscious turkeys, geese, chickens and wild game which they bore, and the buyers and birds alike look well. . . . Turkeys must be had for Thanksgiving and . . . they were." And so, the residents of New York City were spared yet another insult.

Fast-forward to our modern era, with all the advanced molecular genetic techniques and evolutionary models that we have at our power to study the origin and spread of diseases like influenza. In 2014, University of Arizona professor Michael Worobey and his colleagues proposed a tantalizing possible link between the Great Epizootic and the human flu pandemic of 1918, which resulted in an estimated twenty-five to fifty million deaths worldwide. To make sense of that link, we need to understand that based on various evolutionary models, many scientists are now reasonably confident in attributing the Great Epizootic to a strain of equine influenza that is known as equine H7N7 (a strain now thought to be extinct and replaced by the equine H3N8 strain).

Using sophisticated, novel models grounded in molecular biology and evolutionary biology, Worobey and his team set out to understand the genetic origin of the flu strain responsible for the 1918 pandemic. In a paper published in *Nature*, using more than eighty thousand gene sequences of flu strains from birds, horses, humans, pigs, and bats, they found that massive changes—called a "selective sweep" in the literature on evolution—to the genome of avian influenza virus (AIV) began in the late 1800s and continued up through (and then past) the outbreak of the human pandemic in 1918.

Their evolutionary models also found that the equine H7N7 associated with the Great Epizootic, the avian influenza virus that caused henfluenza at that same time, and strains of flu that infect humans are all very closely related to one another: they all share a common-ancestor virus from the mid- to late 1800s. Indeed, these strains are so closely related that Worobey and his team argue that after the selective sweep of the AIV genome was underway, the "AIV lineage subsequently contributed most of the genomic segments to the 1918 pandemic virus." That was a very important finding in and of itself, as there was (and to some extent still is) debate on whether the incredibly virulent form of flu associated with the 1918 pandemic had its origins in swine or birds. But it is a sentence later in the *Nature* paper where the Great Epizootic's relation to the 1918 flu pandemic comes into the story, albeit indirectly: "We speculate that there may have been some epidemiological event possibly associated with the 1872 epizootic," Worobey and his colleagues write, "that precipitated a global sweep of AIV internal genes, which we infer to have commenced around this time."

What that epidemiological event associated with the Great Epizootic was, the authors don't say, but one possibility is that some birds at the time of the Great Epizootic were coinfected with both the avian and equine strains of influenza, and that some sort of transfer of genes from the nasty equine H7N7 strain to the avian strain might have been involved in the selective sweep in the avian influenza virus, which subsequently played a role in the 1918 flu epidemic. If that happened—and that is a *big*, untested, though reasonable enough idea—then some of the genes that brought the Great Epizootic in 1872 may have come back to haunt us, via avian influenza virus, as a contributing factor to the most deadly outbreak of flu ever recorded in humans.[1]

# Acknowledgments

I am in debt to my friends and colleagues Michael Sims, Henry Bloom, and Edward Dolnick for feedback and encouragement as this project unfolded.

Team Dugatkin was of inestimable help. Dana applied her well-honed proofreading skills to each chapter of the book (many times), and Aaron provided insightful comments on draft chapters.

I am a lucky author to have Janice Audet, editorial director of the MIT Press, to take time from her extraordinarily busy schedule to serve as editor for this book. Without her skills and aid, this book would never have come to fruition. Thanks also to Kathleen Caruso and Julia Collins for their fantastic work during the copyediting phase of production.

Finally, I thank 2R, who is most assuredly not a horse.

# Notes

### Introduction

1. "an epizootic [struck] among the horses" and "began by fever," J. Law, "Influenza in Horses," in *Report of the Commissioner of Agriculture for the Year 1872* (Washington, DC: Government Printing House, 1874), 203–248, 205. For more on glanders, see G. Sharrer, "The Great Glanders Epizootic, 1861–1866: A Civil War Legacy," *Agricultural History* 69 (1995): 79–97. For a history of animal epidemics written just before the Great Epizootic, see G. Fleming, *Animal Plagues: Their History, Nature, and Prevention*, vols. 1 and 2 (London: Chapman and Hall, 1871).

### Chapter 1

1. "Our talk has," "We have covered," and "most of us," "The Position of the Horse in Modern Society," *Nation* 15 (1872): 277–278; "machines by means," R. Thurston, *The Animal as a Machine and a Prime Motor, and the Laws of Energetics* (New York: John Wiley and Sons, 1894), 1. For more on the horse as a living machine, see C. McShane and J. Tarr, *The Horse in the City* (Baltimore: Johns Hopkins Press, 2007); J. Tarr and C. McShane, "The Horse as an Urban Technology," *Journal of Urban Technology* 15 (2008): 5–17; and A. N. Greene, *Horses at Work: Harnessing Power in Industrial America* (Cambridge, MA: Harvard University Press, 2008). Whether Descartes denies that animals have feelings and emotions is more complicated; see J. Cottingham, "Brute to the Brutes?: Descartes' Treatment of Animals," *Philosophy* 53 (1978): 551–559; and P. Harrison, "Descartes on Animals," *Philosophical Quarterly* 42 (1992): 219–227.

2. "handsome lady painted," *New York Evening Post*, June 28, 1832; "whirled through the streets," *New York Evening Post*, April 16, 1832; "[This] property is

calculated," *New York Evening Post*, January 17, 1835; "the size of this [new] spacious vehicle," "the space between the seats," and "the floor is level," *New York Evening Post*, August 11, 1831; "very scarcely an omnibus passes," *New York Evening Post*, December 21, 1831; "the character of the . . . drivers," "they race[d] up and down," and "a ferocious spirit," *New York Journal of Commerce*, May 6, 1835.

Reinagle's painting *View of St. Paul's Church and the Broadway Stages, N.Y.* is described in the *New York Evening Post*, September 9, 1831. For more on hackney coaches in New York City, see G. R. Taylor, "The Beginning of Mass Transportation in Urban America, Part I," *Smithsonian Journal of History* (Summer 1966): 35–50. For more on wages in the 1830s, see S. Lebergott, "Wage Trends, 1800–1900," in *Trends in the American Economy in the Nineteenth Century* (Princeton: Princeton University Press, 1960), 449–500. For more on omnibuses, see N. Papayanis, *Horse-Drawn Cabs and Omnibuses in Paris: The Idea of Circulation and the Business of Public Transit* (Baton Rouge: Louisiana State University Press, 1996); McShane and Tarr, *The Horse in the City*; and J. Miller, *Fares, Please! From Horse-Cars to Streamliners* (New York: Appleton, 1941).

3. *The Strangers Hand-book for the City of New York: What to See and How to See It* (New York: C. Francis and Company, 1853), 97–98; "a suite of rooms," *Boston Evening Transcript*, July 12, 1849. Roxbury was annexed to Boston in 1868, West Roxbury in 1874, and Dorchester in 1870, but omnibuses were running between Boston proper and these locations long before that (see *Boston Evening Transcript*, October 24, 1849, for a list of omnibus lines). For more on transportation in Boston, see E. H. Derby, "Commercial Cities and Towns of the United States—No. XXII—Boston," *Hunt's Merchants' Magazine and Commercial Review* 23 (1850): 483–496; G. R. Taylor, "The Beginning of Mass Transportation in Urban America, Part II," *Smithsonian Journal of History* 1, no. 3 (Autumn 1966): 9–52; and S. B. Warner Jr., *Streetcar Suburbs: The Process of Growth in Boston, 1870–1900* (Cambridge, MA: Harvard University Press, 1978).

For more on Chicago omnibuses, see S. Schlickman and L. Klabunde, *The History of the City of Chicago Central Area Transit Circulation Efforts*, NURail Working paper ID: NURail2017-UIC-R17 (2018), https://rosap.ntl.bts.gov/view/dot/61456; B. L. Pierce, *A History of Chicago*, vol. 2 (Chicago: University of Chicago Press, 1976); and D. M. Young, *Chicago Transit: An Illustrated History* (DeKalb: Northern University Illinois Press, 1998).

4. "There is plenty of philosophy," "while the fresh sunlight," and "a new school of philosophers," G. Foster, *New York in Slices/By an Experienced Carver* (New York: W. S. Burgess, 1849), 65; "hirsute, porky creatures" and "stop at the New York Hotel," *The Diary of George Templeton Strong: Post-War Years, 1865–1875*, ed. A. Nevins and M. H. Thomas (New York: Macmillan, 1952), July 3, 1868 entry, 221; "the fluttering morning robes," "upon their various errands," "soon the brokers," and "drawn up in extempore barricades," Foster, *New York in Slices*, 65–66; "perfect bedlam," *New*

*York Herald*, October 2, 1864; "Reason and right," *New York Times*, May 20, 1871; "No regular omnibus-rider," Foster, *New York in Slices*, 63; "Five months," C. Lyell, *A Second Visit to the United States of North America* (London: J. Murray, 1849), 253; "No stint of omnibuses," C. Dickens, *American Notes for General Circulation* (London: Chapman and Hall, 1850), 55; "Great wooden ark[s]," Dickens, *American Notes*, 57.

For more on George Foster, see D. Yotova, "Antebellum Urban Reporting as Literary Journalism and Muckraking: George G. Foster's City Sketches in the New York Press," *Journalism History* 44 (2019): 221–231. For more on Charles Lyell in New York, see G. M. Friedman, "Charles Lyell in New York State," in *Lyell: The Past Is the Key to the Present*, ed. D. J. Blundell and A. Scott, Special Publications 143 (London: Geological Society, 1998), 71–81.

5. "transportation of persons and property," *Laws of the State of New York, Passed at the Fifty-Fourth Session of the Legislature* (Albany, NY: E. Croswell, 1831), 231; "The daily practice of carriages," *Brooklyn Daily Eagle*, July 19, 1855; "The convenience afforded," "patronized by the rich," and "the easterly lines," John Noble, *Facts Respecting Street Railways: The Substance of a Series of Official Reports from the Cities of New York, Brooklyn, Boston, Philadelphia, Baltimore, Providence, Newark, Chicago, Quebec, Montreal and Toronto* (London: P. S. King, 1886), 26; "The conductor came up" and "there was no law against," *New York Times*, June 30, 1864; "The time was set" and "after proceeding a few hundred feet," *Globe*, September 10, 1861.

For net receipts on streetcars, see Alexander Easton, *A Practical Treatise on Street or Horse-Power Railways* (Philadelphia: Crissy & Markley, 1859). For more on the New York and Harlem Railroad and streetcars, see E. Stratton, *The World on Wheels* (New York: Published by the author, 1878); H. J. Carman, *The Street Surface Railway Franchises of New York City* (New York: Columbia University Press, 1919); Miller, *Fares, Please!*; A. Harlow, *The Road of the Century: The Story of the New York Central* (New York: Creative Age Press Inc, 1947); J. H. White, *Horsecars, Cable Cars and Omnibuses* (New York, Dover, 1974); B. J. Cudahy, *A Century of Subways: Celebrating 100 Years of New York's Underground Railways* (New York: Fordham University Press, 2003); J. R. Stevens, "The New York & Harlem Rail Road Companies," *Headlights* 46 (1984): 2–6; R. Post, *Urban Mass Transit* (Westport, CT: Greenwood Press, 2007); T. J. Humphrey, *Hourly Coaches, Omnibuses, and Horse Railroads: A History of Horse-Drawn Transit in Massachusetts* (Published by the author, 2023), accessed October 30, 2025, https://static1.squarespace.com/static/595e970172af65c0eba5db50/t/64a823d69359d17c0f2ea484/1688740829360/Horse-Drawn+Transit+History+2023-01.pdf.

6. For more on Boston streetcars, see *Evidence and Arguments Before the Committee on Horse Railroads of the Legislature of Massachusetts, Session of 1872* (Boston: Wright & Potter, 1872); P. Cummings, *Street Railway System of Boston. Professional and Industrial History of Suffolk County, Massachusetts*, vol. 3 (Boston: Boston History Company, 1894), 286–303; A. E. Pinanski, *The Street Railway System of Metropolitan Boston* (Boston: McGraw Publishing, 1908); *The Charters and Locations of the Several Street*

*Railways in the City of Boston*, vol. 1 (Boston: Mudge and Sons, 1869); Y. Sakakibara, "The Influence of the Introduction of Street Railways upon Urban and Industrial Growth in Eastern Massachusetts, 1855–1875" (Master's thesis, Amherst College, 1956); and Humphrey, *Hourly Coaches*.

7. "was a great [steam] railroad centre," "give the poor man," and "ardently desire[d]," Mayor John Charles Haines's Inaugural Address, 1859, https://www.chipublib.org/mayor-john-charles-haines-inaugural-address-1859/; "having received numerous enquiries," Easton, *A Practical Treatise*, v; "A wet day" and "Not an omnibus is seen," Easton, *A Practical Treatise*, 6–7; "the principle cities," Noble, *Facts Representing Street Railways*; "we take no precautions," "The Position of the Horse in Modern Society," *The Nation* 15 (1872): 277–278.

### Chapter 2

1. Erie Railroad owned 600 stockcars and 3,000-plus stockcars, *New York Herald*, July 10, 1872; another railway reported shipping more than 130,000 (live) cattle and horses, *Wisconsin State Journal*, December 10, 1872; "we are about to become the Metropolis," *Ottawa Citizen*, February 27, 1867; "all the horses of the Street Railway Co." and "the prevailing epidemic," *Ottawa Citizen*, October 19, 1872; "the epizootic amongst the horses," *Ottawa Citizen*, October 19, 1872; "carted to the woods," *Ottawa Citizen*, November 9, 1872; "All these horses at present attacked," *Ottawa Citizen*, November 8, 1872. The *Ottawa Times* issued a chilling warning to readers: "This terrible disease (Epizootic Hippo-Diphiros) may be communicated from the horse to human beings. Persons handling sick horses should be careful of inhaling the breath of diseased horses and must be careful to wash their hands after touching one," *Ottawa Times*, October 22, 1872.

2. "POOR BEASTS," "the pestilence among the horses," "Piles of 'em," and "to accommodate themselves," *New York Herald*, October 24, 1872; "The Disease in Boston," "one of which affects man," "the stable be well ventilated," "the nostrils should be sponged," and "been tried and a gentleman," *Boston Globe*, October 24, 1872; "HARROWING TO THE HORSE," "Our horses are attacked," "The afflicted beast," and "The degree of prostration," *Chicago Evening Post*, October 24, 1872.

### Chapter 3

1. "On Monday, the 30th of September," Andrew Smith to Adoniram Brown Judson, November 20, 1872, in *Public Health: Reports and Papers Presented at the American Public Health Association in the Year 1873* (New York: Hurd and Houghton, 1875), 97.

D. McEachran and A. Smith, *The Canadian Horse and His Diseases* (Toronto: James Campbell, 1867). For a historical overview of Markham, Toronto, and other cities and towns in Ontario, see *Census of Canada* (Ottawa: I. B. Taylor, 1873); and C. P. Mulvany, ed., *History of Toronto and County of York, Ontario,* vols. 1–3 (Toronto: C. B. Robinson, 1885). Symptoms of the catarrhal fever can be found in dozens of newspaper articles, including an early report in the *Hamilton Spectator,* October 7, 1872. For more on Andrew Smith and the Ontario Veterinary College, see C. A. Barker "The Ontario Veterinary College: Temperance Street Era," *Canadian Veterinary Journal* 16 (1975): 319–328; and F. Gattinger, *A Century of Challenge: A History of the Ontario Veterinary College* (Toronto: University of Toronto Press, 1962).

2. "splendid commodious," advertising pamphlet for the Rossin House Hotel, https://www.blogto.com/city/2011/01/toronto_of_the_1870s; *Whose Wife* and "sensation follows fast," Toronto *Globe,* October 28, 1872; *Dick the Detective, Globe,* November 18, 1872; O'Brien's *Four Great Shows* and the Great Golden Menagerie, *Globe,* July 10, 1872; German singer and composer Hermine Küchenmeister-Rudersdorf, *Globe,* October 23, 1872.

3. "An epidemic," *Toronto Leader,* October 5, 1872; "The disease has attacked," *Toronto Mail,* October 7, 1872; "had treated no less than," *Hamilton Spectator,* October 7, 1872 (*Ottawa Citizen,* October 9, 1872); "This distressing disease," *Mail,* October 8, 1872; "two thirds of the horses," *Canada Farmer,* November 15, 1872; "vast hospital for diseased horses," *Perth Courier,* October 11, 1872; new disease names such as hippodiptheria, *Hamilton Spectator,* October 28, 1872; new disease names such as hippolaryngialmorbificality and hyposnoolcalrunnosis, *Hamilton Spectator,* November 26, 1872; "The market has been badly supplied," *Kingston Daily News,* October 7, 1872. For more on streetcars in Toronto, see *First Report of the Toronto Street Railway Company* (Toronto: W. C. Chewett and Printers, 1862); A. H. Sinclair, "The Toronto Street Railway," *The Quarterly Journal of Economics* 6 (1891): 98–105; M. Filey, *Not A One-Horse Town* (Richmond Hill, ON: Firefly Books, 1990); and L. Pursley, *Street Railways of Toronto,* 1861–1921 (Los Angeles: Electric Railway Publications, 1958).

4. "chlorate of potash," "relief may be afforded," and "complex mixtures," A. Smith, "Canadian Horse Distemper," in *Diseases of the Horse, and Cattle and Sheep: Their Treatment with a List and Full Description of the Medicines Employed,* ed. R. McClure (Toronto: A. H. Hovey, 1873), 88; "the horse disease," *Victoria-Warder,* October 9, 1872; "Remedy for the Horse Distemper," *Hamilton Spectator,* October 17, 1872; "is your horse sick,?" *Globe,* October 14, 1872.

On the idea that particles emitted by bad air and bad water could be transmitted from individual to individual, see, for example, D. Diderot and J. d'Alembert, *Encyclopédie,* vol. 10 (Paris: Briasson, David, Le Breton and Durand, 1765), 484–485, entry on "MIASME"; probably depending upon," Smith, "Canadian Horse Distemper," 87; "the miasma era;" P. Thorsheim, *Inventing Pollution: Coal, Smoke, and Culture in Britain Since 1800* (Athens: Ohio University Press, 2006); "impure air" and "[it is of] the greatest importance," Smith, "Canadian Horse Distemper," 87; "animals are not severely affected," Smith, "Canadian Horse Distemper," 89; "eminent veterinarian . . . whose great," *Hamilton Spectator,* October 7, 1872; "We warn our readers," *Brantford Weekly Expositor,* October 18, 1872; "owners of horses" *Hamilton Spectator,* October 14, 1872; "If you wish to save," *Sun,* October 31, 1872.

5. "extortionate prices," *Montréal Herald and Daily Commercial Gazette,* October 17, 1872; "that the epidemic among horses," *Montreal Star,* October 18, 1872; "in consequence of the prostration," *Montreal Herald and Daily Commercial Gazette,* October 16, 1872; "Had [their] plan been adopted," *Montreal Star,* November 5, 1872; "That the epidemic among horses," *Montreal Star,* October 18, 1872; "annually for the investigation," *Montreal Star,* November 26, 1872; "At least two thirds," *Morning Freeman,* October 29, 1872; "It is difficult to get," *Morning Freeman,* October 31, 1872; "spreading with the alarming rapidity," Smith, "Canadian Horse Distemper," 85. For more on the disease in St. John, see L. M. Grandy, "The Era of the Urban Horse: Saint John, New Brunswick, 1871–1901" (Master's thesis, University of New Brunswick, 2004).

6. "there is an epidemic among the horses here," *Detroit Free Press,* October 5, 1872; "coming no one knows how," "Friday morning all the streetcars," "a precautionary measure," "from analogy with similar diseases," and "the call on the drug stores," *Detroit Free Press,* October 25, 1872.

7. "almost all the horses," *Buffalo Commercial,* October 5, 1872; "had become epidemic," *Buffalo Commercial,* October 9, 1872; "has made its appearance," "the effect to purify," *Buffalo Commercial,* October 15, 1872; "It must be in the air," *Buffalo Commercial,* October 25, 1872; "at the present time competent judges," *Buffalo Courier Express,* October 19, 1872; "We pity the poor beasts," *Buffalo Daily Dispatch and Evening Post,* October 21, 1872; "THE CITY ON FOOT," *Buffalo Courier,* October 22, 1872; "There are not enough well horses," *Buffalo Commercial,* October 31, 1872; "public interest in the approaching election," *Buffalo Commercial,* October 29, 1872. For more the city of Buffalo, see *Buffalo City Directory for the Year 1872* (Buffalo: Warren and Johnson, 1872).

8. Four days after horse flu arrives in Rochester, NY, 300 horses were reported ill, *Rochester Democrat and Chronicle,* October 21, 1872; "All branches of business," *Daily Patriot,* October 24, 1872; "public spirited boy" and "The horse is a useful animal," *Democrat and Chronicle,* October 26, 1872.

### Chapter 4

1. For more on Niblo's Garden Theater and the Black Crook, see J. P. Blair, "Productions at Niblo's Garden Theatre, 1849–1862" (PhD diss., University of Georgia, 1982); S. S. Day, "Productions at Niblo's Garden Theater, 1862–1868, During the Management of William Wheatley" (PhD diss., University of Oregon, 1972); and K.

Gintautiene, "The Black Crook: Ballet in the Gilded Age (1866–1876)" (PhD diss., New York University, 1984). For more on New York City as the Great Epizootic appeared, see E. Burrows and M. Wallace, *Gotham* (New York: Oxford University Press, 1990); D. W. Sisson, *The Stranger's Guide to New York City, 1871* (New York: D. W. Sisson, 1871); J. Miller, *Miller's New York As It Is, or Stranger's Guide-Book to the Cities of New York, Brooklyn, and Adjacent Places* (New York: J. Miller, 1872); C. E. Prescott, *The Hotel Guests' Guide to the City of New York* (New York, G. W. Averell, 1872); J. McCabe, *Lights and Shadows of New York Life; or, the Sights and Sensations of a Great City* (New York: National Publishing Company, 1872); and E. Crain, *The Gilded Age in New York, 1870–1910* (New York: Black Dog & Leventhal Publishers, 2016).

2. For more on Tweed and Tammany Hall, see K. Ackerman, *Boss Tweed: The Rise and Fall of the Corrupt Pol Who Conceived the Soul of Modern New York* (New York: Carroll and Graf, 2005); Burrows and Wallace, *Gotham*; T. Golway, *Machine Made: Tammany Hall and the Creation of Modern American Politics* (New York: Liveright, 2014); and A. B. Callow, *The Tweed Ring* (New York: Oxford University Press, 1966). For the crude mortality rate in New York City in 1870, see M. R. Haines, "The Urban Mortality Transition in the United States, 1800–1940," *Annales de Démographie Historique* 101 (2001): 33–64. For more on the history of the United States Postal Service, see R. R. John, *Spreading the News: The American Postal System from Franklin to Morse* (Cambridge, MA: Harvard University Press, 1995). For more on freight and express companies, see A. L. Stimson, *History of the Express Companies and the Origin of American Railroads* (New York: Privately published, 1859).

3. "Great is the fire-horse," A. D. Downes, *Fire Fighters and Their Pets* (New York, Harper and Brothers, 1907), 95; "a box beneath the driver's seat," R. Carlisle, *An Account of Bellevue Hospital* (New York: Society of the Alumni of Bellevue Hospital, 1893), 96; "respond at once" and "Then it must know how," E. Emerson, "Making Policemen of Horses," *Harper's Weekly*, January 30, 1909, 27–28. For more on the fire of 1835, see D. Levy, *Manhattan Phoenix: The Great Fire of 1835 and the Emergence of Modern New York* (New York: Oxford University Press, 2022). For more on the fire of 1845, see *New York Tribune*, July 21, 1845; and *New York Herald*, July 20, 1845. For more on the fire of 1776, see B. Carp, *The Great New York Fire of 1776: A Lost Story of the American Revolution* (New Haven: Yale University Press, 2023). For more on the early years of New York Fire Department, see R. Calhoun, "From Community to Metropolis: Fire Protection in New York City, 1790–1875" (PhD diss., Columbia University, 1973). For more on fire horses, see M. H. Smith, *Sunshine and Shadow in New York* (J. B. Burr and Company, 1869).

4. "men of horse affinities," *New York Herald*, October 24, 1872; the exact count of 38,372 horses in city stables reported in the *Annual Report of the Department of Health of the City of New York*, vol. 1870/1871 (New York: The New York Printing Company, 1871), 651. In "The Great Epizootic of 1872," O. Lazarus relied on a contemporary newspaper account from the *Manchester Guardian* (November 15, 1872) to report

the number of horses in New York City as 70,000, but how the *Manchester Guardian* came up with that figure is unclear, and 70,000 is incorrect; see O. Lazarus, "The Great Epizootic of 1872: Pandemics, Animals, and Capital in 19th Century New York City" (Master's thesis, New York University, 2021). An agriculture compendium of the Ninth Census in 1870 shows that 536,861 horses were living on farms across the state of New York and 319,380 listed as living "not on farms"; see *A Compendium of the Ninth Census, June 1, 1870* (Washington, DC: Government Printing Office, 1872), 697. In 1870, New York City had a population of 942,292; see *Ninth Census, The Statistics of the Population of the United States, from the Original Return of the Ninth Census, June 1, 1870* (Washington, DC: Government Printing Office, 1872).

5. "the stench arising," *The Citizens' Association's Report upon the Sanitary Condition of the City* (New York: D. Appleton, 1865), 264; "A FILTHY CITY," *New York Herald*, May 1, 1872; "a fareful stink," *New York Herald*, July 10, 1872. For more on the horse manure problem and rendering companies, see J. Tarr, "Urban Pollution—Many Long Years Ago," *American Heritage* 22 (1971): 65–106; J. Tarr, *The Search for the Ultimate Sink: Urban Pollution in Historical Perspective* (Akron, OH: University of Akron Press, 1996); M. Kiechle, *Smell Detectives: An Olfactory History of Nineteenth-Century Urban America* (Seattle: University of Washington Press, 2017); M. V. Melosi, *The Sanitary City: Urban Infrastructure in America from Colonial Times to the Present* (Baltimore: Johns Hopkins University Press, 2000); and M. Chieffalo, "Dung, Death, and Disease: Livestock and Capitalist Urbanization from the Early Nineteenth Century to the Present" (PhD diss., Harvard University, 2021).

6. "such a conglomeration of merchandise," "The crowd that is continually passing," and "To enumerate the splendid streamers," *A Peep into Catherine Street or the Mysteries of Shopping*, as cited in. B. Puckett, *The History of Lord & Taylor, 1826–1926* (New York: Guinn, 1927), 8–10; "handsome shawls, rich scarves" and "choice laces," *New York Times*, November 9, 1872. For more on horses and stationary power used to make bricks and much more, see J. Tarr, "A Note on the Urban Horse as a Power Source," *Journal of Urban History* 25 (1999): 434–448; and McShane and Tarr, *The Horse in the City*. For more on the delivery of foodstuffs and more to homes and businesses, see Greene, *Horses at Work*. For more on department stores, see W. Weisman, "Commercial Palaces of New York: 1845–1875," *Art Bulletin* 36 (1954): 285–302; R. Hendrickson, *The Grand Emporiums: The Illustrated History of America's Great Department Stores* (New York: Stein and Day, 1980); and M. Domosh, "Shaping the Commercial City: Retail Districts in Nineteenth-Century New York and Boston," *Annals of the Association of American Geographers* 80 (1990): 268–284.

7. "grand equestrian," *New York Times*, February 23, 1871; "Allowing the average," *Sixth Annual Report of the Board of Commissions of Central Park* (New York: William. C. Bryant, 1863), 31; "as regular as clockwork" and "the beach covered by parties," *Brooklyn Eagle*, June 14, 1870. For more on Central Park, see M. Hecksher, "Creating Central Park," *Metropolitan Museum of Art Bulletin* 65 (2008): 1–74; and for more on

the animals in the Central Park menagerie, see *Board of Commissioners of the Public Parks, Annual Report* (New York: William C. Bryant & Co, 1871). For more on the early years of the Metropolitan Museum of Art, see K. Baetjer, "Buying Pictures for New York: The Founding Purchase of 1871," *Metropolitan Museum Journal* 39 (2004): 161–243.

**Chapter 5**

1. "streetcar companies and livery stables," both *New York Herald* and *New York Tribune*, October 5, 1872; "an epidemic of a catarrhal nature," *New York Times*, October 5, 1872; in the paper's "Minor Items" section, *New York Times*, October 10, 1872; "The epidemic among horses," *New York Times*, October 19, 1872; "The Canadian horse epidemic," *New York Herald*, October 20, 1872; "trac[ing] the track of the pestilence," *New York Herald*, October 23, 1872.

2. "on the evening of October 21st," A. F. Liautard, "Report on the Epizootic as It Appeared in New York," *Annual Report of the Board of Health, New York, for 1872* (New York: D. Appleton, 1873), 276–279; President and Mrs. Grant's visit is covered in the *New York Herald*, October 22, 1872; "Canadian Epidemic Spreading" and "mixed with two," *New York Herald*, October 22, 1872 (also see *New York Herald*, October 22, 1872); "THE EPIZOOTIC," "Seven Thousand Horses," Millions of Dollars," "How the Disease Will Affect," "squalid shed that shelters," "My dear fellow," "It was the same story everywhere," "A fiendish diligence," and "although there are some differences," *New York Herald*, October 23, 1872; "the disease is not infectious," *New York Times*, October 23, 1872 (also see the *New York Sun*, October 23, 1872); "the air is surcharged with this epidemic," *New York Herald*, October 31, 1872; "cosmical agencies operate," *New York Herald*, November 4, 1872; "the almost fatal illness of the Prince," *New York Sun*, January 1, 1872; "a constant flood of foul humidity," *New York Herald*, July 10, 1872; "A dozen different methods of treatment," *New York Times*, October 26, 1872; "The results of this disease" and "A walk from Harlem," *New York Herald*, October 23, 1872; "thoroughfares . . . crowded with weary workers," *New York Sun*, October 31, 1872; "disease will probably be felt," *New York Herald*, October 23, 1872; "You are directed"—this directive appeared in many newspapers, including the *New York Sun*, October 24, 1872; "a large number of South American ponies," *Boston Evening Transcript*, October 30, 1872; "to prevent the horse disease," *New York Herald*, November 2, 1872; "It prevailed to such an extent," A. D. Judson, "Report on the Origin and Progress of the Epizootic Among Horses in 1872," in *Annual Report of the Board of Health, New York, for 1872* (New York: D. Appleton, 1873), 267.

For more on Alexandre Liautard, see L. M. Crawford, "Tribute to Alexandre Liautard, the Father of the American Veterinary Profession," *Journal of the American Veterinary Medical Association* 169 (1976): 35–37; P. de Wailly, "Tribute to Alexandre Liautard," *Journal of the American Veterinary Medical Association* 172 (1978): 528–530;

and F. Smithcors, *The American Veterinary Profession: Its Background and Development* (Ames: Iowa State University Press, 1963).

3. "THE HORSE EPIDEMIC," *New York Herald*, October 24, 1872; "OUR NOBLE HORSES DYING, *New York Sun*, October 24, 1872; "a horsey-looking fellow," "the great trouble in obtaining," "we have 1,000 horses," "Many of the Fire Department Horses," and "It is calculated," *New York Herald*, October 24, 1872; "contagious and infective" and "If it is contagious," October 24th Report of the Board Health, in *New York Tribune*, October 26, 1872; "The chief topic in this city," *New York Sun*, October 24, 1872.

4. "THE NOBLE HORSE IN PERIL," *New York Sun*, October 26, 1872; "THE EPIZOOTY EXTENDING," *New York Herald*, October 26, 1872; "STRETCHED IN THE STRAW," *New York Herald*, November 2, 1872; "PROGRESS OF THE PESTILIENCE," *New York Herald*, October 25, 1872; "do not seem overanxious," *New York Herald*, October 25, 1872; "The bad and cruel," *New York Herald*, October 25, 1872; "There is something heroic," *New York Herald*, October 28, 1872; "No amount of conductorial ingenuity," *New York Herald*, October 26, 1872; "Some well-informed men," *New York Sun*, October 26, 1872; "soon New Yorkers," *New York Herald*, October 26, 1872; "Crowds of poor and ragged men," *New York Herald*, October 27, 1872; "the yo-ho-ho," *New York Herald*, October 31, 1872; "We shall be glad," *New York Herald*, October 28, 1872; "long lines of walking funerals," *Boston Globe*, October 29, 1872; "fear[ed] some derangement in the mails," *New York Herald*, November 4, 1872; "In many of the poorer parts," *New York Sun*, October 29, 1872.

5. "It is only when we are threatened" and "Without him we should," *New York Tribune*, November 1, 1872; "We really ought," *Nation*, October 31, 1872; "The threatened consequences," *New York Tribune*, October 26, 1872; "The freight depots," *New York Times*, October 31, 1872; "Steamship companies are suffering," *Boston Globe*, October 29, 1872; "large sums have been paid" and "Several grain dealers," *New York Sun*, November 1, 1872. For more on Latta and steam dummies, see S. Seidman, "Alexander Bonner 'Moses' Latta, Nineteenth-Century Inventor and Entrepreneur," *Journal of Kentucky Studies* 29 (2012): 140–148.

6. "The dry goods trade," *New York Sun*, October 26, 1872; "The rattle of the milk cart," *Boston Globe*, October 29, 1872; "the attendance was small," *New York Herald*, October 30, 1872; "considerin' the horse disease," *New York Herald*, October 27, 1872; "the two very valuable," *New York Herald*, October 27, 1872; "The hackmen, with characteristic rapacity," *New York Herald*, October 26, 1872; "from 50 to 100%," *New York Sun*, October 26, 1872.

## Chapter 6

1. "MANIFESTATIONS OF LIBERAL EPIZOOTIC" and "a question worth considering," *New York Herald*, October 26, 1872; "The rich may ride in chaises," *New York Herald*, October 26, 1872.

2. "the fetid quality of the atmosphere," *New York Herald*, November 5, 1872; "The 'hospital' branches," *New York Times*, October 24, 1872; "the great horse epidemic," "Solar researches," and "numerous vegetable blights," *New York Herald*, November 5, 1872; "may be caused by noxious exhalations," *New York Herald*, October 24, 1872; "There is at present hardly anything," Andrew White to Ezra Cornell, June 24, 1868, in J. Nolen, "A Great Teacher of Great Men: James Law, America's First University Veterinary Professor Raised the Bar for Academic Standards," *American Veterinary Medical Association News* 242 (2013): 286–289; "diploma mill," J. F. Smithcors, "Medical Men and the Beginnings of Veterinary Medicine in America," *Bulletin of the History of Medicine* 33 (1959): 330–341; "I have looked through," Andrew White to Ezra Cornell, July 2, 1868, in Nolen, "A Great Teacher of Great Men"; "[The] Hon. Ezra Cornell," *New York Times*, July 17, 1868; "would be one of a series of lectures," "technical and not of the highest," and "could not be the result of conditions," *Cleveland Leader and Morning Herald*, November 7, 1872.

For more on John Gamgee, see J. Francis, "John Gamgee (1831–1894): Our Greatest Veterinarian," *British Veterinary Journal* 118 (1962): 430–438. For more on James Law, see Nolen, "A Great Teacher of Great Men"; and J. Chediak, "Shaping the Profession: James Law and the Rise of Professionalism in the U.S.," *Veterinary Heritage: Bulletin of the American Veterinary Medical History Society* 28 (2005): 1–5. For more on the history of veterinary medicine at Cornell University, see E. P. Leonard, *A Cornell Heritage: Veterinary Medicine 1868–1908* (New York: New York State College of Veterinary Medicine, 1979). For more early veterinary programs in the United States, see *Past Veterinary Colleges in United States and Canada*, accessed October 30, 2025, https://www.avmhs.org/past-veterinary-colleges; and *Former Veterinary Institutions in the United States*, accessed October 30, 2025, https://www.avmhs.org/_files/ugd/7e004b_0395862f84024c089b58c5ed66e1ac56.pdf; E. B. Miller, "Private Veterinary Colleges in the United States, 1852–1927," *Journal of the American Veterinary Medical Association* 178 (1981): 583–593; E. H. Davis, "The Boston Veterinary Institute (1855–1860): Its Establishment, Pioneering Founders, Officers and Graduates," *Veterinary Heritage* 38 (2015): 62–83; and A. Liautard, "History and Progress of Veterinary Medicine," *American Veterinary Review* 1 (1877): 5–19.

3. "It was not uncommon," *New York Times*, October 24, 1872; "the hack horses on the stands," *New York Herald*, October 24, 1872; "a tall, elegant figure," *Detroit Free Press*, March 26, 1867; "a cultured swell," B. Unti, "The Quality of Mercy: Organized Animal Protection in the United States 1866–1930" (PhD diss., American University, 2002), 75; "He began there that active interference," *New York Times*, March 13, 1888; "he had been impelled," "a dismal picture of the moral," and "all pronounced the lecture," *New York Herald*, February 12, 1866; "Patrons of a Society" founding document of the ASPCA, in H. Bergh, "Patrons of a Society Proposed to Be Founded for the Prevention of Cruelty to Dumb Animals," 1866, Archives of the ASPCA, New York, NY; "one hundred of the best men" and "a species of Declaration of Independence," in "Henry Bergh's Story: The Friend of the Brute Creation

Visits the Quaker City," September 22, 1884, ASPCA scrapbook, vol. 9: 1882–1887, Archives of the ASPCA; "I know of nothing in nature," "Why should a man," and "persuaded that the obduracy of heart," *American Society for the Prevention of Cruelty to Animals: Fourth Annual Report* (New York: ASPCA, 1870), 2, 5, 6; "an observance of tenderness," *American Society for the Prevention of Cruelty to Animals: First Annual Report* (New York: ASPCA, 1867), 3; "no man shall exercise any," the 1641 Massachusetts Body of Liberties, accessed October 30, 2025, https://oll.libertyfund.org/pages/1641-massachusetts-body-of-liberties; "maliciously, wilfully [sic], and wickedly," Respublica v. Teischer, 1 U.S. 335 (1788); "The police force" and "one half of the fines collected," ASPCA charter approved by New State Legislature, April 10, 1866; "every person who shall maliciously kill," New York Revised Statutes 1829, Title 6, Section 26; "whatever amounts to," *Respublica v. Teischer*; "if any person shall over-drive," "any agent of the American," "offenders found violating," and "American Society for the Prevention of Cruelty to Animals shall cause," New York Revised Statutes 1867, Chapter 375, Sections 1–10. What drove concerns about cruelty to animals, and more generally animal protection, was an amalgam of personal experience, law, religion, education, moral codes, and social standing (advocates tended to be middle-class and upper-class men and those prosecuted for cruelty to animals were often the poor, many of whom were immigrants, who relied on animals for their keep).

For more on Henry Bergh, see E. Freeberg, *A Traitor to His Species: Henry Bergh and the Birth of the Animal Rights Movement* (New York: Basic Books, 2020). For more on Bergh and the "Patrons of a Society" document, see D. Ingram, "The Strange Case of Henry Bergh's 'Declaration of the Rights of Animals,'" *Society & Animals* 30 (2020): 479–491. For more on the animal rights movement, see Unti, "The Quality of Mercy"; D. L. Beers, *For the Prevention of Cruelty: The History and Legacy of Animal Rights Activism in the United States* (Athens: Ohio University Press, 2006); S. J. Pearson, *The Rights of the Defenseless: Protecting Animals and Children in Gilded Age America* (Chicago: University of Chicago Press, 2011); A. Robichaud, *Animal City: The Domestication of America* (Cambridge, MA: Harvard University Press, 2019); and C. Priest, "Enforcing Sympathy: Animal Cruelty Doctrine after the Civil War," *Law & Social Inquiry* 44 (2019): 136–169.

4. "We do not suppose" and "extent and diabolical character," *New York Times*, March 11, 1866; "inciting a mob," ASPCA, *Fourth Annual Report*, 50; "at the inconvenience," *The Advance*, June 16, 1879; "There was a terrible storm" and "The mass meeting," *New York Herald*, May 14, 1868; "nothing improves ruffians" and "malevolent benevolence," *New York Star*, June 26, 1881; "car was unusually crowded" and "the commission of the offense," People v. Tinsdale, 10 Abbott's Prac. Rept. (New) 374 (N.Y. 1868); "There are many people," *New York Times*, March 13, 1888.

5. "Mr. Bergh pointed to the yellow," "a glance at the policeman's baton," and "I have simply done," *New York Herald*, October 30, 1872; "Sir, tomorrow will be the

Sabbath," *New York Herald*, November 3, 1872; "plain violation of the Statute," *New York Times*, January 7, 1873.

6. "there was a very large," "children have rights," "alluded to the case of little Mary Ellen," and "There is a Providence in the affairs of men," *New York Times*, December 29, 1874. For more on the relationship between the ASPCA and SPCC (and other organizations), see Unti, "The Quality of Mercy"; Pearson, *The Rights of the Defenseless*; T. Gilfoyle, *City of Eros: New York City, Prostitution, and the Commercialization of Sex, 1790–1920* (New York: W. W. Norton & Company, 1994); and J. Myers, *Child Protection in America: Past, Present, and Future* (New York: Oxford University Press, 2006).

7. "the interval of gloom," "did not look either," and "sorry beasts," *New York Herald*, November 5, 1872; "Merchants had no difficulty" and "faces of the millionaires, "*New York Times*, November 5, 1872; "There has appeared in the large" and "It seems to affect those animals," *New York Herald*, November 12, 1872; "dull and listless," *New York Sun*, November 13, 1872; "the diseases spoken of," *New York Herald*, November 16, 1872; "Most of the horse owners," *New York Sun*, November 13, 1872; "no perceptible interruption to travel," *New York Herald*, November 15, 1872; "The last phase of," *New York Herald*, November 18, 1872; "the business markets," *New York Herald*, November 19, 1872; "All traces of epizootic have disappeared, *New York Tribune*, November 26, 1872 "alarming stories," *New York Herald*, October 24, 1872.

**Chapter 7**

1. "Upon my sole," from "Lament of a Human Sufferer," *Boston Commercial Bulletin*, November 9, 1872; "there is an epidemic," *Boston Globe*, October 5, 1872; "almost all the horses," *Boston Evening Transcript*, October 5, 1872; "There is scarcely a horse," *Boston Evening Transcript*, October 8, 1872; "Toronto horses are suffering," *Boston Globe*, October 9, 1872; "unusually elegant and elaborate assortment," *Boston Evening Transcript*, October 8, 1872; "Fascinating Performances and Laughable Sketches," *Boston Evening Transcript*, October 17, 1872.

2. "several hundred horses" and "took hold of a very large number," *Boston Evening Transcript*, October 22, 1872; "Its speed has been so rapid," *Boston Globe*, October 25, 1872; "The place of the horse," "imagine the streets stripped," "The something in the air," and "a stagnation of business," *Boston Evening Transcript*, October 24, 1872. For the weather report of October 1872 (and 1871), see *Boston Evening Transcript*, November 2, 1872.

3. "physicians tell us that," *Boston Globe*, October 28, 1872; "has been so rapid," *Boston Globe*, October 25, 1872; "exercise promoter," "boot-maker's friend," and "The Corn Doctor's Blessing," *Boston Globe*, October 29, 1872; "The fears of a

suspension" and "Pedestrians from Cambridge," *Boston Evening Transcript*, October 26, 1872; "atmospheric influence," *Boston Evening Transcript*, October 22, 1872; "sunlight dispels more," *Boston Evening Transcript*, March 19, 1872; "an important point to discover," *Boston Globe*, October 26, 1872.

4. "pull-men" wagons, *Boston Evening Transcript*, October 26, 1872; "equine-imity of temper" and "many men have of late," *Boston Evening Transcript*, October 28, 1872; "One of the most melancholy effects," *Chicago Evening Mail*, November 2, 1872; "the usual proportion of diseased [horses]," "The omnibuses of the Citizen's line," and "the iron horses," *Boston Globe*, October 26; *Frank Leslie's Illustrated Newspaper*, November 16, 1872; "In future years," *Boston Globe*, October 28, 1872; "Behold! How she shines in her beauty," from "The Steam Fire Engine," *Fireman's Herald*, March 9, 1882; "contemplate[d] that should the disease spread," *Boston Evening Transcript*, October 26, 1872; "to obtain volunteers sufficient," *Boston Globe*, October 28, 1872.

For more on the fire horses of Boston, see C. Merwin, *Road, Track and Stable* (Boston: Little, Brown and Co., 1892). For more on steam-powered as well as self-propelled steam-powered fire engines, see W. King, *History of the American Steam Fire-Engine* (Boston: Pinkham Press and Browne, 1896) and G. W. Brown. *The Amoskeag Manufacturing Company of Manchester, N. H.: A History* (Printed by the Amoskeag Manufacturing Company, 1915).

5. "All honor to Henry Bergh," *Boston Evening Transcript*, October 10, 1872; "Brother Henry Bergh," *Boston Globe*, October 31, 1872; "make arrests yourself," Bergh to Angell, April 11, 1868, as in Freeberg, *A Traitor to His Species*; "When the human being looks," *Our Dumb Animals*, October 1874, 2; "we have found our people," *Boston Evening Transcript*, November 2, 1872; "The directors of the Metropolitan Railroad" and "We deem it right," *Boston Evening Transcript*, November 7, 1872; "business continues to suffer," *Boston Globe*, November 1, 1872.

6. "Currents and counter-currents" and "The roar of this Niagara," J. Damrell, *Address Delivered by Capt. John S. Damrell before the Boston Veteran Firemen, February 2, 1886* (Boston: Association of Boston Veteran Firemen, 1886), 16; "the use of gunpowder," *Thirty-Fifth Annual Report of the Boston Fire Department, 1872* (Boston: Rockwell and Church, 1873), 29; "I have never forgiven myself," Damrell, *Address Delivered*, 22; "Louis saw this terror," L. H. Sullivan, *The Autobiography of an Idea* (New York: Press of the American Institute of Architects, 1926), 181; "was delayed from three or five," "the horses of the department," "they were rightly left in the stable or the hospital," "were in general use in the city," and "We cannot but feel," *Report of the Commissioners Appointed to Investigate the Cause and Management of the Great Fire in Boston* (Boston: Rockwell and Church, 1873); "the drivers of horses," *Boston Globe*, October 26, 1872.

For more on the Boston fire, see C. Coffin, *The Story of the Great Fire, Boston, November 9–10, 1872* (Boston: Shepard & Gill, 1872); R. H. Conwell, *History of the*

*Great Fire in Boston, November 9 and 10, 1872* (Boston: B. B. Russell, 1873); F. E. Frothingham, *The Boston Fire, November 9th and 10th, 1872. Its History, Together with the Losses in Detail of Both Real and Personal Estate* (Boston: Lee & Shepard, 1873); H. Murdock, *Letters by a Gentleman in Boston to His Friend in Paris Describing the Great Fire* (Boston: Houghton Mifflin Company, 1909); J. Vahey, *The Epizootic Fire*, City of Boston Archives Fire Commissioner records (0500.016), Box 1, Folder 17, undated; A. Sammarco, *The Great Boston Fire of 1872* (Dover, NH: Arcadia, 1997); C. R. Rosen, *The Limits of Power: Great Fires and the Process of City Growth in America* (New York: Cambridge University Press, 1986); and S. Schorow, *The Great Boston Fire: The Inferno That Nearly Incinerated the City* (Guilford, CT: Globe Pequot Press, 2022).

7. "the horse disease is rapidly disappearing" and "were longest afflicted," *Boston Globe*, November 13, 1872; "In an emergency like that," *Boston Globe*, November 14, 1872; "There is a mistake somewhere," *Boston Globe*, November 20, 1872.

**Chapter 8**

1. "Chicago never does things," *Chicago Tribune*, November 5, 1872; "All was animation," *Chicago Tribune*, October 9, 1872; "The pumping works," *Eleventh Annual Report of the Board of Public Works to the Common Council of the City of Chicago* (Chicago: D & C. H. Blakey, 1876), 63; "The preservation of the good order," order issued by Mayor Mason, October 11, 1871; "Hannibal crossing the Alps," *Lakeside Monthly* xiii (October 1872), 282; "The wonderful activity," *Chicago Tribune*, October 9, 1872. For more on the Great Chicago Fire, see R. Miller, *American Apocalypse: The Great Fire and the Myth of Chicago* (Chicago: University of Chicago Press, 1990); Rosen, *The Limits of Power*; K. Sawislak, *Smoldering City: Chicagoans and the Great Fire, 1871–1874* (Chicago: University of Chicago Press, 1995); and *The Great Fire of Chicago: A Full Account of Its Origin and Progress as Seen by Eye Witnesses* (Indianapolis: Indianapolis Printing and Publishing House, 1871).

2. "Smiting the lowly," from "The Epizootic," *Chicago Tribune*, November 17, 1872; "The first cases," *Report of the Board of Health of the City of Chicago: For the Years 1870, 1871, 1872 and 1873* (Chicago: Bulletin Printing Company, 1874); "orders have been issued" and "The directors of several street railway companies," *Inter Ocean*, October 23, 1872; "Forewarned is forearmed," *Chicago Tribune*, October 22, 1872; "the dangers the pestilential breath," *Chicago Evening Post*, October 24, 1872; "Should the disease prove," *Chicago Tribune*, October 24, 1872; "It is fortunate," *Chicago Evening Mail*, October 24, 1872; "epizootic vapors," *Chicago Tribune*, November 16, 1872; "Death Mounted on a Pale," *Chicago Evening Post*, November 4, 1872; "to refuse to carry horses," *Chicago Evening Post*, October 29, 1872; "The idea that the horse," "Of the 20,000 to 25,000 horses," and "whether the residents," *Inter Ocean*, November 2, 1872; "The Streets Deserted" and "legs that seldom walked," *Chicago Tribune*, November 3, 1872.

3. "Building operations have," *Chicago Tribune*, November 4, 1872; "work is to be commenced," *Chicago Evening Mail*, November 7, 1872; "The wide, open space," *Inter Ocean*, October 30, 1872; "during the interregnum," *Chicago Evening Post*, November 5, 1872; "employ men, boys," *Chicago Evening Mail*, November 7, 1872; "GENTS: it is entirely," *Chicago Evening Post*, November 6, 1872; "this branch of trade has," *Inter Ocean*, November 11, 1872; "the coal yards," *Chicago Tribune*, November 7, 1872; "Nine Bailiffs and three Judges," *Chicago Tribune*, November 19, 1872; "their male friends," *Chicago Evening Post*, November 16, 1872; "Its effects on the former," *Chicago Tribune*, November 9, 1872; "procure a rig anywhere," *Chicago Tribune*, November 10, 1872; "there has been a notable," *Chicago Tribune*, November 18, 1872; "the worst Hor'ce," *Chicago Evening Post*, October 30, 1872.

4. "Passing away," *Chicago Evening Post*, November 10, 1872; "Convalescent: Our Equine Patients Recovering," *Chicago Tribune*, November 13, 1872; "a new bill brimming" and "a roaring extravaganza," *Chicago Tribune*, November 17, 1872; "should go now and laugh," *Chicago Tribune*, November 27, 1872.

**Chapter 9**

1. W. D. Kelly, *Why Colored People in Philadelphia Are Excluded from the Street Cars* (Philadelphia: Benjamin C. Bacon, 1866); W. Still, *Underground Railroad* (self-published, 1872); "a Petition for the Colored People" and "The colored citizens of Philadelphia," *North American* and *United States Gazette*, both on August 31, 1859; "The public highways," *Philadelphia Press*, May 10, 1865; Pennsylvania State Legislature Bill 235 was passed in 1867. For more on discrimination on Philadelphia's streetcars in the 1860s, see P. S. Foner, "The Battle to End Discrimination Against Negroes on Philadelphia Streetcars: (Part I) Background and Beginning of the Battle," *Pennsylvania History: A Journal of Mid-Atlantic Studies* 40 (1973): 260–290; P. S. Foner, "The Battle to End Discrimination Against Negroes on Philadelphia Streetcars: (Part II) The Victory," *Pennsylvania History: A Journal of Mid-Atlantic Studies* 40 (1973): 354–379; G. D. Zylstra, "Whiteness, Freedom, and Technology: The Racial Struggle over Philadelphia's Streetcars, 1859–1867," *Technology and Culture* 52 (2011): 678–670; and F. Speirs, *The Street Railway System of Philadelphia: Its History and Present Condition* (Baltimore: Johns Hopkins Press, 1897).

2. "It begins to be painfully apparent," *Philadelphia Inquirer*, November 1, 1872; "The establishment had been thoroughly" and "it was his opinion that half," *Philadelphia Inquirer*, October 31, 1872; "nearly every railway company," "I have directed," "I wish to avoid," "that so long as a horse will," and "On Tuesday night," *Philadelphia Inquirer*, November 1, 1872; "One thing will doubtless," "Coal, wood and flour" and "truckers who depend upon," *Philadelphia Inquirer*, November 6, 1872. For on the Great Epizootic in Philadelphia, see J. F. Flanagan, "On the Backs of Horses: The Great Epizootic of 1872" (Master's thesis: College of William and Mary, 2011).

3. Newspapers in London, Liverpool, Leeds, Ipswich, Newcastle, Bristol, Chester, Coventry, Sheffield, Derby, and Ormskirk ran stories on the Great Epizootic, i.e. *London Times*, November 14, 1872; *Liverpool Daily Post*, October 31, 1872; *Leeds Mercury*, October 28, 1872; *Ipswich Journal*, October 29, 1872; *Newcastle Weekly Courant*, November 1, 1872; *Daily Bristol Times and Mirror*, November 2, 1872; *Chester Chronicle*, November 2, 1872; *Coventry Standard*, November 2, 1872; *Sheffield and Rotherham Independent Courant*, November 4, 1872; *Derby Mercury*, November 6, 1872; *Ormskirk Advertiser*, November 21, 1872. "three horses said to be affected," *Devon and Exeter Gazette*, November 13, 1872.; "so long as the existing horse disease," *Devon and Exeter Gazette*, November 13, 1872; "though the cry may be 'Wolf!,'" *Coventry Standard*, December 13, 1872.

The *Lancet* reports generally low numbers of horse flu cases in the country, *Pall Mall Gazette*, November 22, 1872; the *British Medical Journal* reports on symptoms seen in Devonshire, *Bath Weekly Chronicle and Herald*, November 7, 1872.

4. "the horse malady," *Charleston Daily News*, November 4, 1872; "The excitement in regard to" and "in order that each and every," *Charleston Daily News*, November 6, 1872; "the mysterious Northern malady," "as is very natural, among animals," "Many of the drivers," and "lack of horses," *Charleston Daily News*, November 7, 1872; "staid and venerable merchants," *Charleston Daily News*, November 8, 1872; "Not a sound was heard on the silent street," line of poem published in the *Charleston Daily News*, November 21, 1872.

5. "the epizootic, or horse disease," *Tennessean*, November 16, 1872; "THE EPIZOOTIC: The Horse Plague Reaches Nashville" and "Owing to the presence," *Union and American*, November 17, 1872; "Hacks are being run," *Union and American*, November 22, 1872; "whereas, it is believed," *Tennessean*, November 21, 1872; "If the present rate," *Union and American*, November 24, 1872; "a larger number of horses," *Tennessean*, November 28, 1872; "it is now generally conceded," *Tennessean*, November 30, 1872.

6. "People who had become unaccustomed" and "Upon reflection, it came," *Times-Picayune*, November 9, 1872; "the majority of the opinions," *New Orleans Republican*, November 20, 1872; "the epizootic or hippophalagia," *New Orleans Republican*, November 23, 1872; "Epizoot is here at last" and "the community must bear," *Times-Picayune*, November 23, 1872; "today we have to record," *Times-Picayune*, November 27, 1872; "with a degree of heroism," *Times-Picayune*, December 4, 1872; "Our river front is so extended," *Times-Picayune*, November 28, 1872; "Cotton languishes on the Levee" and "even music refuses," *Times-Picayune*, December 1, 1872; "Convalescent horses begin," *Times-Picayune*, December 5, 1872; "cut up her mother's toilet cushion," *Times-Picayune*, December 8, 1872. For more on the attempt to pass a quarantine, see *Times-Picayune*, November 13, 1872, November 14, 1872, November 15, 1872, and *Annual Report of the Board of Health to the General Assembly of Louisiana*, December 31, 1872 (New Orleans: The Board of Health, 1873).

As in New York and elsewhere, the streetcars of New Orleans had also been a focal point in the fight for desegregation. Until the spring of 1867, African Americans could only board streetcars branded with a star: these cars constituted about a third of those running routes, and even on starred cars African Americans were often excluded by angry whites who saw no reason to yield. The situation turned very violent as African Americans attempted to board cars with no star, and only diffused after Mayor Edward Heath took matters into his own hands, on May 8, 1867, and went to one of the city's public squares to appeal for an end to the violence between whites and African Americans that spilled over into the streets. Shortly after the mayor's actions, Chief of Police Thomas Adams warned citizens of the city: "No passenger has the right to eject any other passenger, no matter what his color. If he does so, he is liable to arrest for assault, or breach of the peace," New Orleans *Daily Crescent*, May 7, 1867. Not long after that, laws were put into place banning discrimination on streetcars. For more on desegregating New Orleans's streetcars, see R. A. Fischer, "A Pioneer Protest: The New Orleans Street-Car Controversy of 1867," *Journal of Negro History* 53 (1968): 219–233; and H. Laughlin-Stonham, *Streetcars and Protest: From Slavery to Civil Rights: On the Streetcars of New Orleans 1830s–Present* (Liverpool: Liverpool University Press, 2020), 53–82.

7. "Westward the star," *Galveston Daily News*, November 28, 1872; "you do not allow," *Galveston Daily News*, November 28, 1872; "to negotiate for a yoke," *Galveston Daily News*, December 5, 1872; "all packed up," *Galveston Daily News*, December 7, 1872.

8. "What will be the effect," *Nebraska State Journal*, November 21, 1872; "The first horse," *Nebraska State Journal*, December 2, 1872; "The Newsboys Sing," *Nebraska State Journal*, December 12, 1872; "The horses of Wm. Parcell," *Nebraska State Journal*, December 5, 1872; "The 'Epizootic' has sent much consternation," *Nebraska State Journal*, December 6, 1872; "the Epizootic is raging," *West Point Republican*, December 12, 1872; "the [fire] engine horses," *The Omaha Evening Bee*, December 11, 1872.

9. "conveyed by men on foot," *Deseret Evening News*, January 27, 1873; "the hacking cough" and "and the rumbling of wheels," *San Francisco Examiner*, April 23, 1873; "fire department has many sick horses" and "Hackman . . . who are so fortunate," *San Francisco Examiner*, April 23, 1873; "The livery-stables," *Daily Alta*, April 21, 1873; "The horse epizootic," "were afflicted with a terrible distemper," and "chickens coming," *Placer Herald*, March 1, 1873.

Native Americans tended to let their horses run free, as discussed in the *Annual Report of the Commissioner of Indian Affairs for the Year 1872* (Washington, DC: Government Printing Press, 1872) and *Annual Report of the Commissioner of Indian Affairs for the Year 1873* (Washington, DC: Government Printing Press, 1873). The best treatment of the Great Epizootic in the West is T. G. Andrews, "Influenza's Progress: The Great Epizootic Flu of 1872–1873 in the North American West," *Utah Historical Quarterly* 89 (2021): 4–30. For more on the Baker Massacre, see P. Wylie, *Blood on the Marias: The Baker Massacre* (Norman: University of Oklahoma Press, 2016). For more on the Modoc Wars, see R. A. McNally, *The Modoc War: A Story of Genocide at the Dawn of America's Gilded Age* (Lincoln: University of Nebraska Press, 2017). Only in the Oklahoma Territory are there no recorded cases of infected horses. The reason for this is hard to know, and, indeed, it is certainly possible there were horses that were infected, but for any number of reasons, including owners not recognizing the symptoms of horse flu, cases went unreported to the authorities or the newspapers.

**Chapter 10**

1. "practical observations," *Twentieth Annual Report of the Secretary of the Board of Agriculture for 1872* (Boston: W. White, 1873), 145; "under circumstances not yet recognized," *Twentieth Annual Report*, 146; "I do not say that it is contagious," *Twentieth Annual Report*, 153; "We found that all the horses" and "in stables where there are a hundred," *Twentieth Annual Report*, 156; "Well he might have," *Twentieth Annual Report*, 162; "I believe this disease is contagious," *Twentieth Annual Report*, 163; "I mean by contagious disease," *Twentieth Annual Report*, 163–164. For more on Joseph and Edward Lewis Sturtevant, see C. S. Plumb, "Edward Lewis Sturtevant: A Biographical Sketch," *Missouri Botanical Garden Annual Report* (1899): 71–84.

2. Three significant reports on the horse flu epidemic: Liautard, "Report on the Epizootic"; Judson, "Report on the Origin and Progress of the Epizootic"; Law, "Influenza in Horses." In addition there were a few pages written on the epidemic by Canadian veterinarian Andrew Smith that were included in A. Smith, "Canadian Horse Distemper," in *Diseases of the Horse, and Cattle and Sheep: Their Treatment with a List and Full Description of the Medicines Employed*, ed. R. McClure (Toronto: A. H. Hovey, 1873), 85–88. "contagious and infective," *New York Herald*, October 26, 1872; "exposed to unfavorable atmospheric," Liautard, "Report on the Epizootic," 276; "A. F. Liautard, "History and Progress of Veterinary Medicine," *American Veterinary Review* 1 (1877): 5–19.

3. In addition to Judson, "Report on the Origin and Progress of the Epizootic Among Horses in 1872," Judson also wrote a report entitled "Story and Course of the Epizoötic Among Horses upon the North American Continent in 1872–73," in *Public Health: Papers and Reports Presented at the Meeting for the American Public Health Association in the Year 1873* (New York: Hurd and Houghton, 1875); "Proposition No. 1," "The disease prevailed," "in the dry air of Minnesota," and "epizoötic influenza does not spread," Judson, "Report on the Origin and Progress of the Epizootic,"

270; "During the prevalence of the discase," "Investigation has not proved," "the spread of the disease over the country," "The irregularities in time and place," "that the disease appeared," "logically proved," "Epizootic influenza spreads," "Experiments for the demonstration," "Admitting the present impossibility," "No place was exempt from the disease," and "exempt from the disease [they] were so situated," Judson, "Report on the Origin and Progress of the Epizootic," 271; "the disease passed rapidly," Judson, "Report on the Origin and Progress of the Epizootic," 272; "epizootic influenza is the counterpart," "The result of the preceding," "the effects of the prevalence," and "It is evident that the introduction," *Annual Report of the Board of Health*, 291.

4. "the old doctrine," "more than any other epidemic," "the epizootic of 1872," and "The temperature has not exerted," Law, "Influenza in Horses," 218; "sudden *changes* in weather or temperature," "nasal and bronchial catarrhs," "Symptoms exist," and "Overcrowding, with its," Law, "Influenza in Horses," 219; "From these tables," Law, "Influenza in Horses," 224; "remarkably acrid" and "Fogs and vapors," Law, "Influenza in Horses," 228; "Is there a specific contagion," Law, "Influenza in Horses," 233; "If these resulted from atmospheric," "The only theory that will accord," "If it can be introduced," "Were the body merely charged," and "the presence," Law, "Influenza in Horses," 235; "air obtained in stables," Law, "Influenza in Horses," 237; "recognizes in the granules" and "microscopical particles," Law, "Influenza in Horses," 239; "either constituted the virulent principle," "that there is a specific virulent element" and "The air may still be," Law, "Influenza in Horses," 240. Law published a somewhat condensed version of his "Influenza in Horses" in J. Law, "The Causes of Influenza in Horses," *The Lens: A Quarterly Journal of Microscopy and the Allied Natural Sciences*" (January 1873): 1–18. For more on the history of Koch's postulates, see C. Gradmann, "A Spirit of Scientific Rigour: Koch's Postulates in Twentieth-Century Medicine," *Microbes and Infection* 16 (2014): 885–892.

Today we know that influenza is caused by type A, B, or C of the influenza virus. But the first virus to be identified, a tobacco plant virus, was not uncovered until the 1890s, by Russian biologist Dmitry Ivanovsky (1892) and Martinus Beijerinck, a Dutch scientist (1898). It was another four decades before Wilson Smith, C. H. Andrewes, and P. P. Laidlaw, at the National Institute for Medical Research in London, isolated and identified the first flu virus. Today, equine influenza is most often caused by the H3N8 subtype of the influenza A virus (the H7N7 subtype, which also caused equine influenza, is thought to be extinct): the H3N8 subtype was isolated in 1956 and the H7N7 subtype in 1963. Some researchers have suggested that the H7N7 was behind the Great Epizootic, but that remains an open question. In M. Worobey, G. Han, and A. Rambaut, "A Synchronized Global Sweep of the Internal Genes of Modern Avian Influenza Virus," *Nature* 508 (2014): 254–257, the authors suggest H7N7 or something closely related to it caused the Great Epizootic, but Chambers argues that such a position is "conjectural" (T. Chambers,

"Equine Influenza," *Cold Spring Harbor Perspectives in Medicine* 12 [2022], https://doi.org/10.1101/cshperspect.a038331).

**Chapter 11**

1. "There is a discomforting probability," *Chicago Evening Post*, November 22, 1872; "We are now for the first time," *Nation*, October 31, 1872; "it is quite possible," *New York Tribune*, November 1, 1872; "necessity of some solid," *New York Herald*, October 27, 1982; "Humanity is slowly," *New York Herald*, November 3, 1872; "disorganized everything," "some wise cynic has remarked," "depend[ed] . . . upon the animal," "The animal is only a means," and "Let him [the horse] be considered," *New York Times*, July 24, 1881.

2. "a select committee" and "public notice [be] given, *Report of a Special Commission Designated by the Senate to Ascertain the Best Means for the Transportation of Passengers in the City of New York*, New York Public Library Digital Collections, January 31, 1867, accessed January 31, 2025, https://digitalcollections.nypl.org/items/510d47dc-9a8c-a3d9-e040-e00a18064a99; "this enterprise, which has now," *New York Times*, November 6, 1869; "This transit experiment," *Brooklyn Eagle*, November 15, 1870; "ended its precarious existence," *New York Times*, August 25, 1870; "There is in New York," *Appletons' Journal of Literature, Science and Art* 65, no. 3 (June 25, 1870). For more on elevated rails in New York, see R. Reed, *The New York Elevated* (New York: A. S. Barnes, 1978); E. Oszustowicz, *The Elevated Railways of Manhattan* (New York: Electric Railroaders Association, 2016); C. Cheape, *Moving the Masses: Urban Public Transit in New York, Boston, and Philadelphia, 1880–1912* (Cambridge, MA: Harvard University Press, 1980); J. Nevius, "The Elevated Era," June 27, 2018, accessed January 31, 2025, https://ny.curbed.com/2018/6/27/17507424/new-york-city-elevated-train-history-transportation; B. Horbal, "The Early Proposed Railways for New York City, Part 1," January 16, 2015, accessed January 31, 2025, https://www.nypl.org/blog/2015/01/16/early-proposed-railways-nyc-1; and B. Horbal, "The Early Proposed Railways for New York City, Part 2," July 7, 2015, accessed January 31, 2025, https://www.nypl.org/blog/2015/07/07/early-proposed-railways-nyc-2.

3. "The defeat of the Underground Railroad," *New York Times*, April 11, 1864; "Though the horse," *New York Times*, June 10, 1894; "First Car Off the Earth," *Boston Globe*, September 1, 1897; "monumental buildings, electric vehicles," A. N. Greene, "Harnessing Power: Industrializing the Horse in Nineteenth Century America" (PhD thesis, University of Pennsylvania, 2004).

The notion of pneumatically driven transportation would, on occasion, raise its head, including in 1870, when Rufus Gilbert, using some of the technology that Beach had developed, proposed a pneumatic-powered elevated railway. The design and technology were exquisite—two parallel running pneumatic tubes, suspended twenty-four feet above ground by arches cast in Gothic style, with pneumatic-powered elevators at each station—but the funding was lacking. Pneumatic power was also being considering for the long-distance transportation of freight. In March 1872, a few months before the Great Epizootic was wreaking havoc and commerce was slowing to a crawl, the U.S. Senate was considering a remarkable, if somewhat bizarre bill. Bill S. 883, if passed, would have incorporated the National Pneumatic Tube Company, with the aim to construct a pneumatic tube system from New York City to New Orleans; freight would be packed into hollow balls and blown through tubes from one city to the other. Congress thought better of it and Bill S. 883 died on the floor.

For more on pneumatic railways, see *Illustrated Description of the Broadway Underground Railway, with Explanatory Details and Engravings* (New York: Beach Pneumatic Transit Co., 1872), Rare Book Division, New York Public Library Digital Collections, accessed February 10, 2025, https://digitalcollections.nypl.org/items/510d47dc-9aab-a3d9-e040-e00a18064a99; K. H. Courage, "The First Subway in New York City Was a Cylindrical Car Pushed by Air," *Scientific American*, September 1, 2020, https://www.scientificamerican.com/article/the-first-subway-in-new-york-city-was-a-cylindrical-car-pushed-by-air/; S. Łotysz, "Alfred Beach Not Alone: The American Patents of Pneumatic Railway in 19th Century," *Icon* 9 (2003): 93–107; and G. Goldin and S. Lubell, *Never Built New York* (New York: Metropolis Book, 2016). For more on early subways in Boston and New York, see D. Most, *The Race Underground: Boston, New York, and the Incredible Rivalry That Built America's First Subway* (New York: St. Martin's Press, 2014); C. Hood, *722 Miles: The Building of the Subways and How They Transformed New York* (New York: Simon & Schuster, 1993); B. Clarke, *Tremont Street Subway: A Century of Public Service* (Boston: Boston Street Railway Association, 1997); and Cudahy, *A Century of Subways*. For more general information on mass transit in this era, see Cheape, *Moving the Masses*; B. Cudahy, *Cash, Tokens, and Transfers: A History of Urban Mass Transit in North America* (New York: Fordham University Press, 1990); Miller, *Fares, Please!*; and Post, *Urban Mass Transit*.

4. "Perhaps we shall have it," *Times-Picayune*, November 9, 1872; "an unqualified success" and "the cars moved along," *Chicago Tribune*, January 29, 1882; "Electricity had taken such a hold," F. Sprague, "Some Personal Experiences," *Street Railway Journal* 24 (1904): 566; "the road presented conditions," Sprague, "Some Personal Experiences," 570; "They came, of course," O. W. Holmes, "The Broomstick Train," *Atlantic Monthly* 66 (1890): 247–248. For more on cable cars in Chicago, see G. Borzo, *Chicago Cable Cars* (Mount Pleasant, SC: Arcadia, 2012).

## Epilogue

1. "we have bad news for" and "appears to resemble the [horse]," *New York Herald*, November 18, 1872; "dying by the thousands," *Nashville Union and American*, November 19, 1872; "the hooks of the stalls fairly," *New York Herald*, November 28, 1872; "AIV lineage subsequently contributed" and "We speculate that there may

have been," Worobey, Han, and Rambaut, "A Synchronized Global Sweep," 254, 256. For more on horse, avian, and human influenza, see D. Morens and J. Taubenberger, "An Avian Outbreak Associated with Panzootic Equine Influenza in 1872: An Early Example of Highly Pathogenic Avian Influenza?," *Influenza and Other Respiratory Viruses* 4 (2010): 373–377; T. Chambers, "Equine Influenza," *Cold Spring Harbor Perspectives in Medicine* 12 (2022): a038331, https://doi.org/10.1101/cshperspect.a038331.

# Index